AF545039

Fluid dynamic measurements in the industrial and medical environments

Volume 2 Discussion

Edited by David J. Cockrell

Leicester University Press

Fluid dynamic measurements in the industrial and medical environments

Fluid dynamic measurements in the industrial and medical environments

Proceedings of the Disa Conference
held at the University of Leicester
April 1972

Volume 2 Discussion

Edited by David J. Cockrell
Senior Lecturer in Engineering
University of Leicester

Leicester University Press 1973

First published in 1973 by Leicester University Press
Distributed in North America by Humanities Press Inc., New York

ISBN 0 7185 1114 X

Printed in Great Britain by Unwin Brothers Ltd, Old Woking, Surrey

Contents

Preface
David J. Cockrell

In June 1970, nearly two and a half years ago, planning began for the Conference on Fluid Dynamic Measurements in the Industrial and Medical Environments, the Disa Conference. Now, months later and a successful Conference past, it is interesting to look at what our committee wished to achieve and try to evaluate its achievement.

Our main objective was to seek to gather together those who were concerned with experimental fluid mechanics in other than idealised situations. The proper attitude for the experimentalist in "pure" research we felt was to make measurements in an environment specially designed to ensure the dependence of the measured quantity on as few parameters as possible. But in industry or in medical research it seemed that such isolation would seldom be realised. We envisaged situations in which fluid properties would present special difficulties and others where there might exist geometrical or scale problems so that measurements, though necessary, would be extremely difficult to make. The common ground between the industrial and the medical environments would be, in our view, the relative remoteness of this much more readily available pure research.

Whilst the existence of a medical environment in which fluid dynamic measurements are made is clear, our concept of an industrial environment is much more open to question. The work in the medical environment is performed by research workers from a number of institutions, most of which are connected with hospitals and universities. There is no obvious parallel for the industrial environment, a fact which was increasingly brought home to us as we tried to select appropriate material for the Conference. The industrially-based fluid dynamicist is as concerned as anyone else in, for example, the use of optical techniques to measure high turbulence intensities though he may have less direct opportunity to assist with their development than others in academic and research institutions. In the early stages of receiving papers for consideration we were disappointed at the preponderance of those from academics but later we appreciated that many of these papers, if only their authors could state clearly the practical applications of their work, discussed the backbone of fluid dynamic research in the industrial environment.

Dr. Sovran has much more to say about the nature of the industrial environment in his review lecture printed in Volume I. I believe that he was like ourselves in anticipating that he clearly appreciated the nature of the field we had selected for him. Nevertheless I know he found it hard to define clearly the field's boundary conditions and my committee are particularly grateful for the thought he gave and the matter he presented in his published paper.

The second problem over which we had given much anxious thought was the bringing together of workers in these two dissimilar environments. Though there are some who like Dr. Schultz can move freely from the industrial to the medical fields and back again they are in an obvious minority. Would the concern in largely common instrumentation lead to interest in one another's research fields and would there be positive benefit in bringing the two groups together?

Early in our planning we had hoped to amalgamate the two fields completely both at the Conference and in its Proceedings. A number of incidental factors made this impracticable including an important medical conference which overlapped the first day of our own. Inevitably therefore some industrial fluid dynamicists left the Conference early and some medical fluid dynamicists arrived late. However, common ground was found and perhaps most significantly in Dr. Schultz's excellent review lecture, published in Volume I of the Conference proceedings.

It is not given to Conference organisers to measure the success of their Conference, theirs is only the relief that all has passed without the onset of too many more grey hairs. There were disappointments, notably the difficulty experienced by some overseas contributors to obtain the funds necessary to be with us in person. Others have kindly expressed their appreciation and the contents of this particular volume, which include Mr. Peter Bradshaw's valuable summary on the papers dealing with hot-wire and hot-film anemometry measurements, indicate the liveliness of all those who participated.

I have already thanked my fellow committee members in Volume I. Here I want to express the thanks of all those associated with the Conference to the Disa Company. The conception of the Conference was theirs, most of the secretarial burden was carried by them and yet their presence throughout was restrained and dignified. At the Conference Dinner their Managing Director announced a grant to be made by the company to further fluid dynamic research and details of the Disa Grant are given later in this volume.

Other contents of Volume II include summaries of the questions raised and the statements made by Conference participants Questions were referred to the authors of the Volume I papers and here they have given their considered replies. For the purpose of easy reference the lay-out and numbering of the papers in Volume II is the same as in Volume I.

Where no reply to a question has been received from an author the question does not appear in this volume.

The Conference made a small financial profit. It was the wish of the Conference Committee and the Disa Company that this be given to the University of Leicester in return for the hospitality provided in April 1972.

David J. Cockrell.
December 1972.

Measuring mean and fluctuating velocities by hot-wire and hot-film techniques

Rapporteurs comments at the Disa Conference by P. Bradshaw

The theme of this section is the calibration of hot-wire and hot-film anemometers, and the use of those calibrations to derive velocity components from voltage measurements. The papers overlap a good deal - more in aim than in content - and it is convenient to discuss the following topics rather than the individual papers.

1. Dependence of voltage on velocity for a wire normal to a steady stream.

2. Dependence of effective cooling velocity on yaw angle.

3. Dependence of velocity sensitivity on fluid temperature.

4. Frequency response of the probe-plus-anemometer system.

5. Measurement of mean and fluctuating velocity in highly turbulent flow.

6. Digital processing of fluctuating voltages (especially with application to (5)).

At the time of writing I had not seen the papers in other sections: I apologise for (but do not necessarily withdraw) any remarks which contradict what other authors have said.

1. DEPENDENCE OF VOLTAGE ON VELOCITY FOR A WIRE NORMAL TO A STEADY STREAM

The variation of Nusselt number with Reynolds number has been studied by several authors recently and it has been suggested that the simple power-law form

$$Nu = A_1 + B_1(Re)^n \tag{1a}$$

$$\text{or} \quad E^2 = A + B\,U^n \tag{1b}$$

with n = 0.5 according to the slightly dubious arguments of King or 0.45 according to the painstaking experiments of Collis and Williams, is not adequate. A few general points can be made: some of them are made in the other papers, notably Paper II 4-3 by T. Davies and Patrick, and Paper II 4-6 by Bruun and P. Davies.

(i) There is no reason why the Nusselt number <u>should</u> follow a simple power law: the only question is whether this or any other law is an adequate approximation over the Reynolds number range required in a given experiment.

(ii) Collis and Williams' experiments (1)(which I for one am inclined to trust) were done with wires of high aspect ratio (more than 2000) at low Mach numbers (less than 0.14). Typical anemometer wires have aspect ratios of the order of 200: the heat loss to the supports is significant, and it alters significantly with air speed. Also, whether or not compressibility effects as such are apparent, the Mach number may have a significant effect on the Nusselt number at given Reynolds number (2) because M/Re is proportional to (mean free path)/(wire diameter) and typical anemometer wires operate on the verge of slip-flow effects.

(iii) It is very difficult to be certain that a curve fitted to experimental points has the right gradient everywhere (especially at the ends of the range). Perry and Morrison (3), de Haan (4) and others have suggested that dynamic calibration (i.e. direct measurement of the gradient of the voltage/velocity curve) is advisable. Unfortunately Perry and Morrison's paper can be interpreted as implying that static calibration is inherently wrong (rather than merely difficult to do to sufficient accuracy). I don't think they really mean this and I certainly don't believe it. However, dynamic calibration has much to recommend it especially for linearized wires.

Apart from recourse to dynamic calibration, the following proposals have been made for improving on eqn.(1) and are discussed in Paper II 4-6 by Bruun and P. Davies.

(a) use of power law with n a function of speed.
(b) replacement of $(Re)^n$ by a tabulated function derived from many measurements with probes of similar type (A and B being chosen to fit a particular wire, as usual).
(c) addition of a term [C Re] to the power law formula with n = 0.5 (this idea is due to Siddall and T. Davies of Sheffield: it is described by T. Davies and Patrick in Paper II 4-3 and used by Dvorak and Syred in Paper II 4-1).

If data are to be reduced on a computer a tabulated function (or a piecewise-analytic one) is quite acceptable but for hand calculations a single analytic function is preferable: Bruun and P. Davies' results confirm T. Davies and Patrick's demonstration that the latter's "one-and-one-half" power law is an excellent fit to the former's tabulated function, and Dvorak and Syred show that the "one-and-one-half" law closely fits Perry and Morrison's dynamic calibration results. The law

$$E^2 = A + B\sqrt{U} + C\,U \qquad (2)$$

is a little difficult to fit by hand, or to invert by an analogue linearizer, but one imagines that an adequate approximation would be

$$E^2 = A + B\sqrt{U}\,(1 + C_1\sqrt{U}) \qquad (3)$$

where C_1 is taken as known from previous results (-0.02, with U in metres/sec, from Davies and Patrick's fit to the ISVR data). I did some further analysis and found that the Davies and Patrick formula can be approximated by a 0.45 power law with an accuracy of better than 3 percent on slope over the range 1 to 30 metres/sec which is a larger range than is likely to be covered in any one low-speed tunnel experiment. At higher speeds a 0.45 power law is clearly unsatisfactory. Bruun and P. Davies obtain similar results. Perhaps one may tentatively conclude that the Collis and Williams 0.45 power law is adequate at low Mach numbers, even for short wires, but that for speeds above about 40 metres/sec. the Davies and Patrick analytic fit or the Bruun-Davies tabulated function should be used: one would like to see more use of dynamic calibration, at least for the establishment of even better fits or tabulations. One final comment is that Bruun and Davies point out that the zero-speed voltage (determined by free convection) is not simply related to the constant A of a simple power law fit to the forced-convection heat transfer: it is not clear to me why this objection does not apply also to the Bruun-Davies tabulated function (certainly T. Davies and Patrick reject the zero-velocity points).

2. DEPENDENCE OF EFFECTIVE COOLING VELOCITY ON YAW ANGLE

There are as many opinions (but fewer data) on yaw sensitivity as on velocity sensitivity. Of the authors in this section Dvorak and Syred (Paper II 4-1), Durst and Rodi (Paper II 4-5) and Thomas, Brown and Birch (Paper II 4-8) assume that the effective cooling velocity U_{eff} is given by

$$U^2_{eff} = U^2 \, (\cos^2\alpha + k^2 \sin^2\alpha) \qquad (4)$$

where α is the angle between U and the plane normal to the wire ($\alpha = 0$ for a conventional single-wire probe aligned with the stream) Dvorak and Syred show that k varies greatly with speed for a typical Disa probe; Durst and Rodi and Thomas et al do not discuss the value of k. Bruun and P. Davies assume

$$U_{eff} = U \cos^m \alpha \qquad (5)$$

but do not discuss the value of m. Since k or m is obtained from $(\partial E/\partial V)/(\partial E/\partial U)$ its value is sensitive to the velocity calibration from which $\partial E/\partial U$ is obtained. It is also very sensitive to the measurement of α: personally I doubt whether the angle between the wire and the airstream can be measured to better than about 1 deg. and for α near 45deg this implies an uncertainty of 3% in yaw sensitivity; in the Aero Department at IC (5) we assume $U_{eff} = U\cos\alpha_{eff}$ and calibrate the wires individually to find the effective angle α_{eff}.

Cheesewright (Paper II 4-2) points out that none of the simple laws is valid for $\alpha > 70$ deg. The other authors who are concerned with high-intensity turbulence do not seem to have considered this point.

3. DEPENDENCE OF VELOCITY SENSITIVITY ON FLUID TEMPERATURE

Cheesewright mentions the large amount of calibration data needed if the fluid temperature, as well as the velocity and yaw angle, is variable. Elsner (Paper II 4-4) presents some useful comparisons of "theoretical" velocity and temperature sensitivities, derived from Kramers' (6) and from Collis and Williams' calibrations, with his own experimental results (note that in Elsner's Fig. 3 the Collis data are plotted right to left). The differences in temperature sensitivity are quite large: in particular the experimental variation of temperature sensitivity with fluid temperature is much larger than predicted by either "theoretical" expression. Elsner is probably right in attributing this to departures from two-dimensionality - i.e. the effects of heat loss to the supports.

4. FREQUENCY RESPONSE OF THE PROBE-PLUS-ANEMOMETER SYSTEM

Noskievic (Paper II 4-7) presents evidence of yet another eccentricity of hot film probes. In addition to the Bellhouse-Rasmussen effect of a fall in sensitivity from the zero-frequency value to a lower plateau at a few hundred Hertz, he finds that, when operated at constant temperature by Disa 55A01 or 55D01 anenometers, Disa sensors exhibit a decrease in sensitivity as the frequency is reduced below 30Hz. Noskievic indicates difficulties in the use of sinusoidal voltage excitation to simulate fluctuating heat transfer and an alternative demonstration is the discrepancy (Fig.2 of Paper II 4-7) between hot-film measurements and measurements by the magnetohydrodynamic method described by Janalik in Paper II 5-1.Quite clearly something is wrong somewhere, and I hope that the technical staff of Disa will be able to comment on the suggestion that the electronics, rather than the probe, is responsible.

5. MEASUREMENT OF MEAN AND FLUCTUATING VELOCITY IN HIGHLY TURBULENT FLOW

The two full-length papers, II 4-1 by Dvorak and Syred and II 4-2 by Cheesewright, both treat the problems of measurements in high-intensity flows. Dvorak and Syred consider the extraction of three mean velocity components and six Reynolds stresses from measurements with a single wire set successively at three different angles to the flow (certain covariances being obtained separately). In the main part of their analysis they work in terms of the effective cooling velocity Z - most simply described as the output of a linearized wire. The velocity components are expressed as rather complicated functions of the instantaneous **Z** (using eqn.(4) for yaw response) which are then expanded as Taylor series: first and second derivatives with respect to Z appear in the evaluation of mean velocity, and are retained in the fluctuating terms. Dvorak and Syred do not discuss the likely error of their approximation in flows with large fluctuations in direction. They use eqn.(4) for yaw response and as a consequence it is most convenient to obtain, say, $\overline{u^2}$ as $\overline{(U + u)^2}-U^2$.

I have outlined Dvorak and Syred's approach in advance of their presentation to compare it with Durst and Rodi's (Paper II 4-5). The latter authors deal with two-dimensional mean flows $U_3 = 0$, $\overline{u_1u_3} = 0$, $\overline{u_2u_3} = 0$. In this case, again using eqn.5 for yaw response, the mean square of the effective (mean plus fluctuating) cooling velocity can be expressed as a rather more tractable function of the four non-zero components of $U_iU_j + \overline{u_iu_j}$ and of the wire angle; $U_iU_j + \overline{u_iu_j}$ can therefore be extracted from measurements of the mean-square output of a linearized wire set successively at four different angles. Unfortunately the determination of the mean velocity components is left as an exercise to the student! Once they are found, the Reynolds stresses $\overline{u_iu_j}$ can be extracted from $U_iU_j + \overline{u_iu_j}$:the process is of course well-conditioned only if the turbulence intensity is high.

Thomas, Brown and Birch (Paper II 4-8) also deal with high-intensity flow, using a two-wire probe set successively at two different angles. Again mean squares and mean products of the effective cooling velocity are used in the analysis but in order to estimate the unmeasured covariances between the fluctuating voltages measured at successive positions, it is assumed, following T. W. Davies, that the voltages have a square waveform with constant phase. The plausibility of this assumption is not discussed: it implies that the skewness of the voltage fluctuation is zero and that certain covariances are unity, which may be far from true sometimes.

Cheesewright emphasises the point that if the angle between the instantaneous flow angle and the wire axis is less than about 20 deg. (for which the probability exceeds 0.1 if $\sqrt{\overline{u^2}}/U$ exceeds 0.3 and α = 45 deg.) the simple yaw calibration laws fail: the variation of wire voltage with yaw angle in this region is (i) complicated (ii) very sensitive to probe geometry because of the effect of the prong wakes. This is probably the main obstacle to measurements in highly-turbulent flows.

Bruun and Davies show that for $\sqrt{\overline{u^2}}/U$ less than about 0.3 the difference between linearized and non-linearized measurements of $\overline{u^2}$ is negligible: this emphatically does not apply to triple-product or spectrum measurements, and Cheesewright shows that errors in $\overline{v^2}$ or $\overline{uv}$ can be significant.

6. DIGITAL PROCESSING OF FLUCTUATING VOLTAGES (ESPECIALLY WITH APPLICATION TO (5))

Digital processing of the instantaneous voltage signal is used by several authors. Its main advantages are that complicated statistical quantities can be evaluated more easily than by analogue means, and that complicated calibration or linearization curves can be used. Cheesewright's paper is in itself a rapporteur's comment on the subject, and is particularly valuable because he actually admits that computer time costs money.

REFERENCES

(1) D.C.Collis and M.J.Williams "Two-dimensional convection from heated wires at low Reynolds numbers", J.Fluid Mech., Vol.6, p.357, 1959.

(2) F.W.Boltz "Hot-wire heat loss characteristics and anemometry in subsonic continuum and slip flow", NASA TN D-773, 1961.

(3) A.E.Perry and G.L.Morrison "Static and dynamic calibrations of constant-temperature hot-wire systems", J.Fluid Mech., Vol.47, p.765, 1971.

(4) R.E.de Haan "The dynamic calibration of a hot wire by means of a sound wave, Appl.Sci.Res., Vol.24, p.335, 1971.

(5) P.Bradshaw "An Introduction to Turbulence and its Measurement" Pergamon 1971.

(6) see S. Corrsin "Turbulence: Experimental Methods" in Handbuch der Physik Vol. 8, pt.2, Springer 1963.

II Measurements of velocity in the industrial environment

II.1 MEAN VELOCITY MEASUREMENTS BY THERMISTOR AND HOT-WIRE TECHNIQUES

II.1-1 Questions addressed to U.K. SINGH and R. SHAW, University of Liverpool

H.H.BRUUN asks: This investigation has been carried out in a turbulent boundary layer. It is well known that local turbulence intensity affects the mean velocity measurements. How does this contribution change the data presented?

I.WILLIAMS asks: Could the authors comment on the causes of the error induced in anemometer probe measurements by the proximity of a pipe wall enclosing fully developed turbulent air flow? In particular would they comment on the effects of wall temperature, varying flow velocity and variations in the probe wire-to-wall distance.

F.MAK asks: What was the length-to-diameter ratio of the hot wires used for your measurements? Did you correct for end losses when comparing the results with those obtained by Wills in reference (2) of your paper?

E.J.BUUR asks: Fig.2 shows an irregular decrease of the velocity at about 9 m/s. Can you explain this?

U.K.SINGH and R.SHAW reply: All four comments on the paper relate to the subdivision of the total error into separate error effects. The present investigation did not seek such a sub-division and the comparison with Wills' correlation was not intended as more than a guide to orders of magnitude. Nevertheless it is intended to study the separate contributions to the error in a further experiment and in reply to I. Williams the authors list these separate effects as:

1. The heat loss to the wall which acts as a large heat sink. This loss will be dependent on the difference in temperature between the hot-wire and the wall (200°C compared with 20°C in the experiments reported in the paper). It will also be dependent on the flow velocities and on the ratio of the distance of the wire from the wall (b) and the wire radius (a) as shown by F. Mak in paper II. 1-3. The velocities involved in the reported experiments would appear to be too high for this effect to be significant even at the minimum distance of 0.12mm when $^{b}/a = 30$.

2. The heat lost via the needles which support the wire. This loss will be dependent on the inclination of the

needles to the flow direction, being greatest when the needles are perpendicular to the flow and least when they are parallel to the flow (reference 3 of the paper) and dependent on the velocity distribution along the needles. In the reported experiments the needles curve from an inclination of 2-deg at the wire to 90-deg to the flow outside the wall shear layer. As stated in F. Mak's comment, this loss is also dependent on the length/diameter ratio of the wire which was 120 in the experiments. The ratio in Wills' experiment was 5 times this value, but no attempt was made to separate this particular loss from the others.

3. The error in the calculated velocity due to the use of steady calibration laws when the turbulent velocity components are large compared with the mean velocity. This is the error referred to by H.H. Bruun who has shown* that this error is positive except when $\bar{u}^2 >> \bar{v}^2$ & $\bar{w}^2$. A positive error in the mean velocity of about 1% (or 0.1m/s) could be expected in the present experiments just outside the viscous sub-layer where the measured turbulence was about 16%. This is apparently the irregularity in the error curve (fig. 2) which E.J. Buur noticed at about 9m/s velocity.

* Linearization and hot wire anemometry, H.H. Bruun, Journal of Physics E: Scientific Instruments, 1971, Volume 4.

II.1-2 Questions addressed to A.ARROWSMITH and P.J.FOSTER, University of Sheffield

F. MAK asks: The effect of heat transfer distortions created by the proximity of your hot-wire to the water drops must obviously vary considerably with drop size. Do you feel that such an error might be significant in your measurements?

F. DURST asks: The presented measurements were performed in a boundary layer of variable thermodynamic properties since the air near the falling droplets is undoubtedly saturated with water vapour. Has the saturation been accounted for in the hot-wire calibration? If not, could the authors assess the possible maximum error?

A.ARROWSMITH and P.J.FOSTER reply: We consider that the effects of the variation of the physical properties of air with relative humidity or any effect of the proximity of the drop stream on the probe calibration was small compared with the uncertainty as to the exact effect of the non-uniform air velocity on the anemometer signal. During calibration the probe support was perpendicular to the flow, the same as in use, and measurements were started some distance from the drop stream in order to reduce such effects.

We have now completed a theoretical analysis which permits a more accurate interpretation of this type of measurement.

II.1-3

Questions addressed to F.J. MAK, The Gas Council, London

A.ARROWSMITH asks: Would Mr. Mak describe the analysis used to justify the assumption that velocity variation along the hot wire does not lead to significant error in velocity measurement?

S.C.KACKER asks: What are the limitations on the scaling up factor for the burner port?

P.A.O.L.DAVIES asks: Did you correct the new velocity measurements you presented for wall effects? Could you give details of the approach you employed for this?

F.J.MAK replies:To A.A.ARROWSMITH, the justification is two-fold. Firstly, Betchov's effective cold length, in conjunction with Wills' heat-transfer data was used to calculate the "sensitive" lengths of the wires considered. The analysis is described in detail in my paper. Secondly, our experimental results showed that the heat loss curves for a 5μm, 1.1 mm wire, obtained near a plane wall, can be used to correct for measurements made near a curved wall accurately, as explained in the text and figures (2) and (3) of my paper.

To S.C.KACKER it has been pointed out in my paper that combustion and aerodynamic measurements have to be made on the same burner port.

With methane as the reference gas, the maximum burner diameter that can sustain a laminar flame between the upper inflammability limit (gas lean) and the stoichiometric limit is about 1 cm*. Increasing the port diameter means that only very lean flames can be obtained; decreasing the port diameter creates further measurement problems.

To P.A.O.L.DAVIES,yes,the figure below shows the results of velocity traverses made across a 1 cm diameter duct of length to diameter

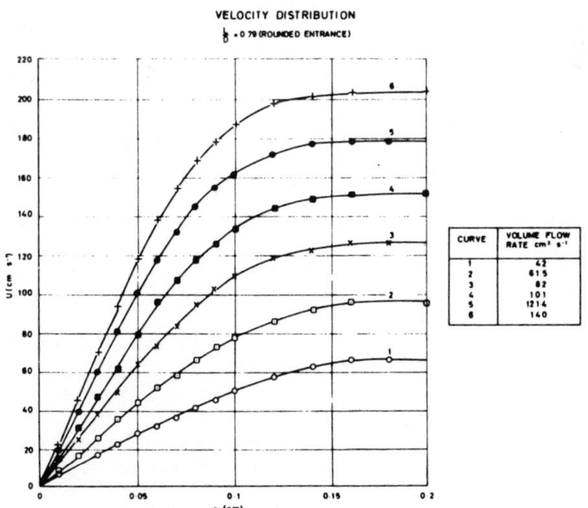

*Combustion, Flames and Explosions of Gases; Lewis and von Elbe, Academic Press Inc., 1961, p.254.

ratio 0.79 with a well rounded inlet section. The velocity points in the wall region (≃1 mm away from duct wall) were corrected using a series of correction curves found from the near-wall measurements in the axisymmetric fully-developed laminar-flow channel, on a relative voltage basis.

The approach employed above was a more direct and practical method, the basis of which is similar to that shown in figure (1) of my paper and discussed in the text.

II.1-5 Questions addressed to Z. SMOLSKI,
Institute of Applied Mechanics,
Warsaw Technical University, Poland.

J.NOSKIEVIC asks: What accuracy of measurement do you expect from the fluid drip expenditure gauge (FDEG)?

J.KIELBASA asks: How does the temperature variation of the gas affect this accuracy?

E.J.BUUR asks: Please quote this accuracy of measurement in both laminar and turbulent flows.

Z.SMOLSKI replies: The accuracy of the FDEG system depends on calibration and on any differences between the conditions of calibration and measurement. We know that by reversing the physical phenomenon here used, we can calibrate the hot-wire probes in a laminar gas flow, e.g. the Hagen-Poiseuille flow, by measuring the flow-rate of the fluid replaced by gas. So this accuracy may be relatively high. For most applications I think 5% will be sufficient.

When the temperature of the gas varies, the electrical signal from the measuring system is not much affected by it since there are at least two sensors in the different arms of the bridge and compensation will occur.

The FDEG system is designed for small flow-rates of fluid (gas) for which the Reynolds Number, even when related to the diameter of the jet, is very low. So there will be laminar flow only.

As has been said in Proceedings Part I, the FDEG system is based on a fluidic element; it has recently been subjected to further investigations. A four jet fluidic element has been manufactured in which the four sensors form a full Wheatstone bridge. This more sensitive than a two-jet element. Two micron platinum Wollaston wire which was previously used has been replaced by 7.63 micron wire since the thinner wire proved too weak and liable to breaking. Possibly tungsten wire would be the most suitable.

Fig. 1 presents the characteristics of a four-jet element: the variable parameter is the current flowing through the sensors. It can be seen that the effect of overheating is considerable for larger values of expenditure. Both for two-jet and for four-jet elements there are zones of insensitivity close to "O" where the calibration points are scattered in a random way.

The latest design of the fluidic element is shown in Fig. 2. There is only one jet in it, and opposite each end of the jet there is a sensor. So a single fluidic element contains one jet and two sensors forming a half-bridge. We can also use two fluidic elements for a full bridge measuring system. The characteristic feature of this solution is the swirling of the stream of the agent before it enters the jet. This may diminish the hydraulic resistance of the jet. This new design is now being thoroughly tested. Its prototype is composed of two syringes fitted together so that the outlets meet and form a single jet. The metal parts of the syringes provide the electrical connections for the sensors placed inside.

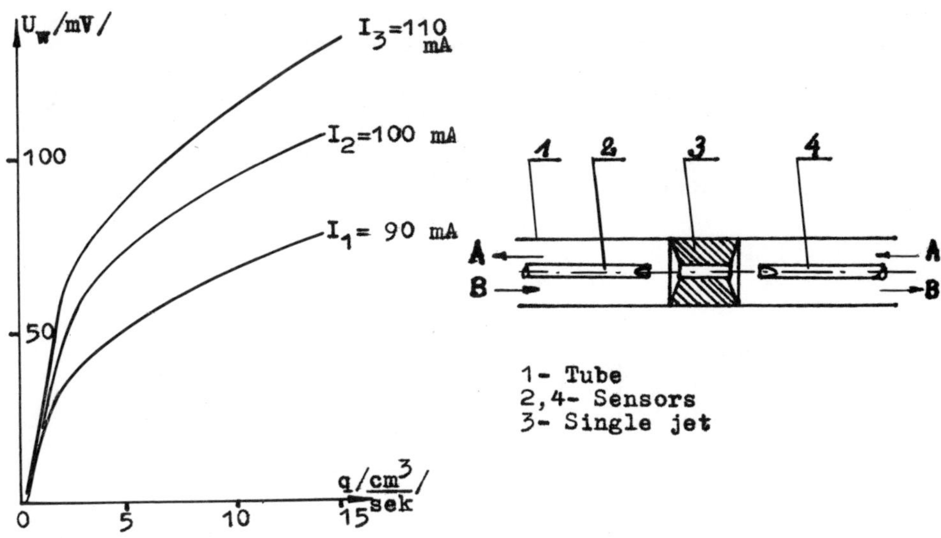

Fig. 1 Calibration curves

Fig. 2 Diagram of New Hot-Wire Fluidic Element

II.1-6 Questions addressed to J. KIELBASA, J. RYSZ, A.Z.SMOLARSKI, B.STASICKI, Applied Physics Laboratory, Polish Academy of Sciences

F.DURST asks: The oscillatory anemometer described seems to perform well in laminar flows and in flows of low turbulence intensity. Do the authors expect similar performance in highly turbulent flows and in separated regions? What is the time resolution with which the instrument can follow velocity variations?

E.A.SPENCER asks: Could further details be given of the development of the authors' new instrument referred to in their presentation? Could they indicate the amount of improvement in sensitivity at various points over the velocity range of this new instrument?

B.STASICKI replies: The oscillatory anemometer works well when the turbulence intensity is sufficiently small, because it is possible to send a temperature signal from the transmitter to the detector without any breaks. If the turbulence intensity is greater, there are short breaks in the oscillations, seen in Fig.1. They cause errors in the frequency meter which works as the pulse counter.

The lowest frequency of oscillations occurs when the flow velocity is very small (0 to 5 cm/s), and equals about 20 Hz. In this case the time resolution is the longest, but it is shorter than 0.1 second.

The indications in the range from 0.5 to 30 m/s are independent of gas temperature and chemical constitution. It is necessary to take into consideration the gas thermal diffusivity when its velocity is lower than 50 cm/s.

The probes ought to be calibrated individually because of differences in spacing between the wires and in wire diameters.

Our oscillatory anemometer is not as sensitive to probe dustiness as other hot-wire anemometers.

The anemometer output quantity is frequency, which can easily be transmitted great distances.

Fig. 1.

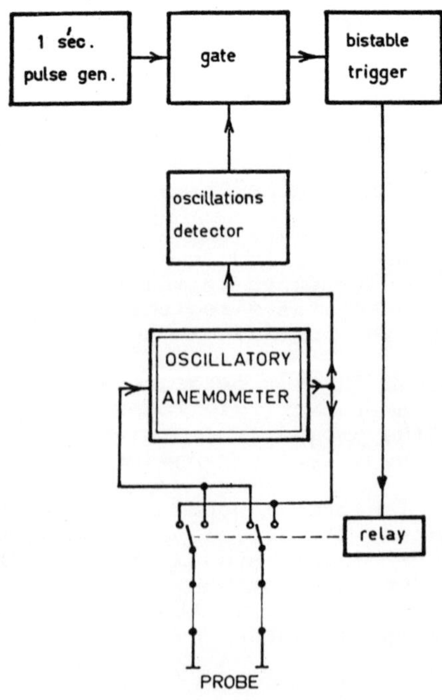

Fig. 2.

Reversal of flow direction causes breaks in oscillation, as the transmitted signal cannot then reach the detector. For flows with reversing directions we can apply the bi-directional oscillatory anemometer shown in Fig. 2.

In this figure, the relay supplied by the pulse generator through the gate and trigger periodically changes the connections to the probe wires. When the wire connected to the output of the anemometer is situated on the windward side, the anemometer begins to oscillate. Then the oscillations detector closes the gate which breaks the trigger and relay connections. Hence the circuit "has identified" the direction of flow. The time of identification when the direction reverses is lower than 1 s.

II.2 MEAN AND FLUCTUATING VELOCITY MEASUREMENTS BY OPTICAL TECHNIQUES

II.2-3 Questions addressed to W.K. GEORGE, Pennsylvania State University, U.S.A.

I.WILLIAMS asks: Could you explain the effect of polydisperse particles on optical measurements in turbulent gas flow, in particular the effect of slip between the larger particles and the fluid?

J.E.RIZZO asks: Your discussion on the break in phase as particles cross the boundary of the sensing region appears to apply only to the fringe field anemometer. In the Doppler instruments is not the light scattered from all particles with the same velocity locked in phase by the wave fronts of the laser beam, depending only on the relative path length through the instrument?

In addition in a real case there are several velocities present. Thus the various discrete frequencies will beat with each other, giving large amplitude variation and drop out below noise level lasting many cycles.

In order to cope with this effect, tracking filters such as the one produced by Disa take a variable frequency band pass centred on the incoming signal and balance the amplitude of frequencies above, against those below. As ambiguity signal is wide-band noise it appears that it does not affect such instruments. Would you comment on this?

W.K.GEORGE replies: To I.A.WILLIAMS, the problem of slip between suspended particles and surrounding fluid have been the subject of many experimental and theoretical investigations.

The first class of criteria that must be considered impose an upper limit on particle size and density. Briefly these are:

(i) the particle size must be smaller than the smallest scale of the turbulence (say the Kolmogorov microscale).

(ii) the particle response time (response to step change in velocity) must be smaller than the smallest time scale of the turbulence (say the time dissipation scale).

In water these criteria are easily met; in air the second criterion is usually not satisfied by many naturally occurring particles.

The second class of criteria impose restrictions on minimum particle size. These include such effects as Brownian motion and thermal diffusion.

In situations where one is forced to use a polydisperse collection of scattering particles, the larger particles would appear to merit the most concern since they contribute disportionately to the Doppler current thus biasing the velocity signal toward the larger particles. If these large particles do not track the flow, the measurement error can be quite significant.

To J.E.RIZZO, with regard to the first comment regarding the applicability of my results to the Doppler instrument let me first state categorically that the (so-called) fringe anemometer and the Doppler anemometer are one and the same - both depending on the Doppler shifted scattered light. The fringes merely form a convenient means of visualizing the scattering volume in some circumstances.

The Doppler ambiguity (in the simplest case of uniform steady flow) is a consequence of the fact that as the population of scatterers changes, the signal loses coherence - the rate of loss of coherence depending on the inverse transit time of particles through the volume. This phenomenon is clearly unavoidable in any optical configuration.

The fact that the Doppler ambiguity is wideband is irrelevant. The point is that the random phase fluctuations are indistinguishable from velocity fluctuations over the band of interest. In addition, the spectral height of the ambiguity is in many cases as high as that of the turbulence. Certainly the mean square ambiguity can be reduced by adjusting the time constant of the tracking loop(on equivalently, by low pass filtering the output). It is also clear that care must be exercised to avoid removing from the spectra some of the turbulence as well.

II.2-6 Questions addressed to R.K. TURTON, <u>Loughborough University of Technology.</u>

L.A. SALAMI asks: What results have been obtained on this interesting project since the paper was submitted for publication?

R.K. TURTON replies: The system has been proved using a rotating disc and is now applied to the pump. Work on the pump has been only possible recently, but there are no results yet due to difficulty in getting a signal that is meaningful. We believe that the trouble is due to spurious scattering at the two water-liquid interfaces and are at the moment realigning the receiving section of the optical path to look along the bisector of the two incident beams. Thus we hope to obtain a cleaner signal for analysis.

II.3 SPECIAL PROBLEMS OF SENSOR CALIBRATION

II.3-2 Questions addressed to T.B. MORROW, Loughborough University of Technology.

S.WOLFENDEN asks: Using untreated water did you experience any bubble formation on the tip of your hot film probe? Using an overheat ratio of 1.1 with the probe in still water were natural convection currents observed? What is the optimum overheat ratio at which to run a hot film probe in the speed range of 0-2ft sec?

T.B.MORROW replies: A detailed account of my own experience with the operation of hot-wire and hot-film sensors in water is contained in reference (3) of the conference paper.

In short, the water was de-aerated before use; when bubbles form on a hot-film sensor, it was often a sign that the electrical insulation has ruptured. I made no observations in still water except measurements of electronic noise. Overheat selection depends upon many factors including the temperature and composition of the liquid, the amount of power available from the anemometer, the desired signal to noise ratio and the temperature and velocity sensitivities. Under the conditions reported in reference (3), an overheat of 25°C was found to work well.

J.T.TURNER says: This paper describes an interesting attempt to produce a correction procedure for the steady flow calibration drift due to dirt accretion on hot wire and film sensors Probably air flow studies are even more troublesome than those using liquid flows in this respect and probe contamination is a problem which has plagued almost all workers in the field. It has certainly been of concern to those based in Manchester (e.g. Collis (1)). The writer would like to offer some of his own experiences with hot wires in air for discussion.

If the steady flow calibration curve (voltage versus velocity) is plotted at successive time intervals, the effect of dirt accumulation will be seen as a reduction in the sensitivity of the system at every velocity level. This results from the conduction barrier around the wire surface. An investigation by Lightning(2), in which electron microscope photographs of the dirt collected around hot wire probes were obtained, has shown that the build-up is never uniform but tends to concentrate on the upstream and downstream faces. The assumption of a mean film thickness is accordingly not realistic except where rough corrections are desired.

However, the prediction of variations in the steady flow calibration is not the only criterion. The standard square wave test for frequency response will show that the dirt film reduces the upper frequency limit of the anemometer system, thus introducing the possibility of erroneous turbulence measurements. Although this is perhaps less serious in liquid flows where frequencies are generally lower, the need for caution is obvious.

These are, therefore, practical reasons why correction for the effects of dirt accumulation is unlikely to be satisfactory. The tendency of the dirt film to break away at irregular intervals, leading to spurious changes in anemometer output, implies that analyses of the problem might be better not attempted. The solution lies in the elimination of contaminants from the flow by careful design of the apparatus. Since this is impossible in most industrial situations, the development of probes which are insensitive to dirt is important. It may be worth recording that increase in size (of the wire) reduces the rate of drift but brings about a reduction in frequency response due to the increased thermal inertia of the system. It may therefore be possible to choose an optimum wire size.

References.

1. Collis, D.C., Ph.D. thesis, University of Manchester, 1959. Also, with Williams M.J.,; Journal of Fluid Mech. Vol.6, p. 357 (1959).

2. Lightning, G., Ph.D. thesis, University of Manchester, to be submitted.

T.B.MORROW adds: An analysis of the effect of sensor contamination on the anemometer output voltage similar to that presented in this paper is given in a discussion of reference (2) by Khader, Elango and Rao, ASCE, Vol. 97, No. HY1, Jan. 1971, pp.194-196.

Sajben has shown that it is possible to calibrate in liquid mercury tungsten hot-wires that have a thin enamel coating and a layer of contamination on the surface of the wire. He includes a term in the energy balance equation that accounts for the conduction heat transfer through both the enamel coating and the contamination layer. The calibration technique is described in "Hot-Wire Anemometer in Liquid Mercury" by M. Sajben, Rev.Sec.Insts., Vol.36, No.7, July 1965, pp.945-949.

II.3-4 Questions addressed to R.W. JOHNSON, Whirlpool Corporation, Michigan and A.M. DHANAK, Michigan State University, U.S.A.

P.O.A.L.DAVIES asks: I realise the difficulties of calibration at very low flow speeds described here and I am puzzled by the results on figure 4 below 0.1 ft/sec. Were the velocity readings very steady for these measurements or not?

The results are not consistent with our experience (see Davies M.R., Graduation Thesis University of Southampton 1968), with measurements using Karman vortex streets behind fine wires, also with results reported elsewhere here (see e.g. fig. 4 paper II.1-3). At low velocities flows tend to separate very easily near the probe so there is always a probability the local velocity was near zero on average.

H.H.BRUUN asks: In correlating the calibration data the following relationship was used: $E - E_o = UT_o/T$. Would you enlarge on the physical or experimental reason for choosing this relationship?

R.CHEESEWRIGHT asks: Would the authors say if they had been able to verify that they had achieved a parabolic velocity profile under all conditions? Our experience suggests that this is not possible.

R.W. JOHNSON and A.M. DHANAK reply: To P.O.A.L.DAVIES, the velocities as indicated both by the flowmeter and the wire were quite steady for all the reported data.

We are not familiar with the details of your work and, therefore, cannot be certain of the reason for any lack of consistency between our observations. One possibility is that there may have been a difference in orientation between the wire and free stream flow in our two setups. As discussed in our reference (6), the behaviour at very low velocities may be explained by the interaction between free convection heat transfer and forced convection heat transfer. Since the free convection heat transfer is influenced by wire orientation, the nature of the interaction may be different for different geometries. However, we would expect that ambiguities will usually be minimized when a relatively low wire temperature is used.

To H.H. BRUUN, quite frankly, the relationship

$$E - E_o = UT_o/T$$

was chosen to correlate our calibration data simply because it was observed to give reasonably consistent results. Some time was spent in attempting to derive a similar relationship using a heat transfer analysis. However, these efforts did not succeed and were abandoned.

To R.CHEESEWRIGHT, we did not verify the existence of a parabolic profile under all conditions. However, a good deal of care was given in the design of the apparatus to ensure the existence of a known velocity profile under all test conditions. In those cases where we expected to have an appreciable deviation from a parabolic profile, results from reference (4) were used to compensate for this expected deviation.

II.4 MEAN AND FLUCTUATING VELOCITY MEASUREMENTS BY HOT-WIRE AND HOT-FILM TECHNIQUES

II.4-1 Questions addressed to K.DVORAK and N.SYRED, University of Sheffield

R.CHEESEWRIGHT asks: Had the authors of these papers plotted their turbulent fluctuation data normalised against the local velocity rather than a mean velocity and thus realised the very high levels of turbulence which they had apparently measured?

W.RODI asks: As a consequence of the truncation of the Taylor series in equation (13), the method neglects third and higher order correlations in the evaluation equations for the Reynolds stresses. Conventional methods involve similar approximations which are questionable in highly turbulent flows. Why then do the authors believe their method to be superior to conventional methods in such flows?

It seems doubtful to me that all three velocity and six Reynolds stress components can be obtained from measurements with a single wire placed under three angles in one plane. Have the authors in fact obtained results for the radial mean and fluctuating velocities?

The shear stress profile seems unrealistic, particularly so near the jet axis, but also at the position of maximum shear; for, together with the mean velocity from Fig. 4, does it satisfy the momentum equation?

The authors state that, in reference (2), the third velocity component was neglected in the derivation of the spatial response equations. Is this statement correct?

D.DURAO asks: Do you have any special reason to explain why in figure 7 of your paper, you present a non-zero shear stress at the centre line?

H.H.BRUUN asks: The analysis introduced in this paper concerning equations(4) to (11) is related to the relationship between the instantaneous values of the effective velocities $Z_1(t)$, $Z_2(t)$, $Z_3(t)$ and the three component velocities U(t), V(t), W(t) as would be measured if the three wires were placed at the same time. The final equations (9) to (11) therefore for each time t contain information with regard to both amplitude and phase between $Z_1(t)$ $Z_2(t)$ and $Z_3(t)$.

In placing wires one at a time the only type of information which will be available will be of the type $\overline{Z_1}$, $\overline{Z_2}$, $\overline{Z_1 Z_2}$ etc.

Would the authors explain how the phase information which is of particular importance for correlation measurements is still retained?

The main reason for the reasonable agreement for mean velocities and turbulence intensity measurements is probably the dependence of these quantities more on amplitude than on phase. For correlation evaluation the method is likely to introduce large errors.

N.E.A.WIRASINGHE asks: Equations (4) and (6) are only sensitive to G and K. In a real flow situation w could be much greater than u, when these equations would be identical. What sort of values of G and K do you get for yaw and pitch angles varying between 0-deg. and 90-deg?

L.A.SALAMI asks: The two papers II.4-1 and II.4-6 claim to have developed a method suitable for high turbulent flow even with non-zero mean velocities in any arbitrary system of co-ordinate of axes x y z. However, the proving of these methods was carried out in jets or channels where the mean velocities in the y and z axes could have been zero.

In industry there are many situations where complex types of flow really exist: e.g., after bends, partly opened valves etc., and it should be possible to justify these new methods by traversing in such flows and comparing the integral with respect to y of the mean axial (x axis) velocities with other methods of flow measurements; e.g. a calibrated orifice plate. Would the authors comment on this procedure?

K.DVORAK and N.SYRED reply: To R.CHEESEWRIGHT, it is commonly accepted that in regions of high turbulence intensities in the free jet the local velocity fluctuations are related to the mean velocity on the centre line - references (2) and (7) in our paper. Normalizing by the local mean velocity tends to disguise the absolute value of turbulent fluctuations and also the changing nature of these quantities through a flow field.

To W. RODI, the Taylor series expansions of the functions given in equations (9), (10) and (11) are truncated at second order terms. We therefore retain second order derivatives of these mean velocity functions and second order correlations in the form of their products. As is pointed out in our paper the velocity function U (Z_1, Z_2, Z_3) is assumed to be a continuously differentiable function whose derivatives are finite at a given point. With this assumption third order derivatives of the mean velocity functions are certainly an order of magnitude smaller than the second order terms. It can be shown that expansions of the Taylor series to the third order results in an expression which contains the mutual combination of third order derivatives (small) and triple correlation products. We do not claim that triple correlation products are negligible but their mutual combination with third order derivatives is small compared to our second order terms and mean velocity. Conventional methods neglect second order terms completely. W. Rodi, reference (2) in our paper, neglects triple correlation products which are not multiplied by a small third order derivative. Hence we feel justified that our method contains less restrictive assumptions. Our method is also operationally fast and convenient to use as all measurements are taken in one plane.

Referring to our paper it can be seen that we take nine measurements in one plane, three mean voltages, three fluctuating voltages and three correlation coefficients.

As pointed out in our paper the accuracy of our method is unavoidably decreased in the region of steep mean velocity gradients where third order terms in the Taylor series expansion may become significant. This is particularly reflected in the expression for Reynolds stress. The other source of error is felt to lie with the unsatisfactory correlator available.

In reference (2) Mr. Rodi neglects the mean tangential velocity and only takes the fluctuating tangential velocity into account ($\overline{U}_3$, u_3 in his notation).

To D.DURAO, the two factors in the penultimate paragraph above probably account for erroneous values of the shear stress on the axis of the jet. The analysis of Reynolds stresses showed that $\overline{u'w'}$ was sometimes significant compared with $\overline{u'v'}$ and therefore the momentum equations should be constructed and satisfied in three directions.

To H.H.BRUUN, for all techniques which do not measure the three velocity vectors instantaneously the assumption must be made that the random fluctuations are stationary in their nature. i.e. their time history is repeatable at any time interval *. With this assumption phase information may be obtained at any interval of time.

We would agree with Dr. Bruun that this assumption does not affect mean velocities and turbulence intensity measurements too much; however, shear stresses are very sensitive to correlation information. Therefore for certain flows where the above assumptions are not satisfactory, instantaneous recording of signals is required for shear stress measurement. For many industrial purposes it is only required to measure mean velocities and turbulence intensities.

*Bendat, J.S., Piersol, A.G. "Measurement and Analysis of Random Data", J. Wiley and Sons, 1966.

To N.E.A. WIRASINGHE, the value of G has been shown to vary little with pitch between $0 < \theta < 90^{\circ}$ especially for DISA type 55F11 gold-plate wire probes. (G is very nearly equal to 1) - reference (5) in our paper. The K factor varies considerably with yaw angle and velocity. The variation with velocity is dealt with in our paper, whilst it is shown in reference (5) that K only varies between 0.1 and 0.3 for yaw angles between 0 and 90°. K^2 always appears in equations (4) to (11) and is therefore at least an order of magnitude smaller than G^2. Hence neglecting the variation of K with yaw angle is justified.

To L.A.SALAMI, we are at present applying our method to swirling flows where all three velocity components are significant. Some success has been achieved.

As comment on P. BRADHSAW'S relevant rapporteurs comments (p. 9 of this volume), K.DVORAK and N.SYRED add: P. Bradshaw suggests that triple correlation products should appear in equations (15) and (17) of our paper in order to maintain the accuracy of the method to the second order of the Taylor series.

Although triple correlation products do not appear in equation (15) for normal stress $\overline{U'^2}$, the accuracy of this expression is higher than to first order as is explained in the following paragraphs.

Theoretical expressions for the variance of functions of random variables including second order terms indeed contain these triple correlations (e.g. Sveshnikov, ref. 6, p. 137) - consequently such expressions have little pratical significance. We have avoided this obstacle by deriving equation (15) in a slightly less rigorous way. Considering U^2 and designate it by a new function $H(Z_1, Z_2, Z_3)$ we calculate its expectation by applying equation (13) (bearing in mind H is a compound function and its second partial derivatives must be calculated correspondingly). Subtracting the square of equation (13) from the expectation of U^2 gives the final expression in equation (15). The whole procedure is rather lengthy but follows logically. Thus the final formula for $\overline{U'^2}$, equation (15), contains second order derivatives; the accuracy is probably lessened slightly by the double application of equation (15). We have found that in the absence of steep main velocity gradients the second order terms have a convergent character and add more accuracy to the technique. Similar comments apply for the other normal and shear stresses.

II. 4-2 Questions addressed to R. CHEESEWRIGHT, Queen Mary College, London

JOAN WHAMOND asks: Don't you consider that for most work rather than use a complex algebraic approximation or a tabulated array and digital techniques it would be just as satisfactory to carry out conversion of voltage to flow velocity by use of an analog function generator?

D.E.M.TAYLOR asks: To increase pay off has a mixed system approach been considered, using analog, for calibration data and digital for the arithmetic? It would be possible to do much of preliminary processing in a complex flow, including statistics, if a parallel hybrid rather than a simple analog computer were used.

E.J.BUUR asks: The other day we decided in our Turbomachines laboratory of TNO-IWECO in DELFT, Holland to purchase a data-acquisition system.

Would you give us more detail than there is in your paper about the Q.M.C. digital system?

R.CHEESEWRIGHT replies: To JOAN WHAMOND, the use of an analog function generator is a possible approach for single wire systems and its justification or otherwise would depend on how closely one wanted to approach the maximum possible level of accuracy, which is attainable only with digital techniques. For two and three wire systems where the voltage to velocity conversion is a 2 to 2 (or 3 to 3) mapping the technique is not really applicable.

To D.E.M.TAYLOR, there are many applications of hot-wire anemometry for which digital processing is not justified. It is only when the velocity range is very large, and fluctuation levels are large that it is likely to be justified. We at Queen Mary College use the approach because we feel we have a flow (natural convection) which justifies it. I have not personally experienced any cases in which a half analog, half digital approach would be advantageous because such an approach would loose many of the advantages of the digital approach as noted in my paper. It is however time that we costed a system using a rather larger mini-computer than the PDP-8L. On such a machine the majority of our processing could be done either in real time or by using a disc store. Such a system might cost £20,000 to £25,000.

To E.J.BUUR, the Q.M.C. digital system and the associated computer programmes are described in detail in this reference* which is available on request from the author.

*G.Hall; "The PDP-8L/T5000 Magnetic Tape Data Logging System"; Queen Mary College, Faculty of Engineering Research Paper No. EP 5001, Dec. 1971.

Discussion between P.BRADSHAW and R.CHEESEWRIGHT concerning digital interpretation of hot-wire data, specifically linearisation of signals. R.CHEESEWRIGHT asked if this meant the recovery of values of velocity from values of voltage, in which case the term was not really appropriate, for in the analog connotation it only meant half of this process. If the whole process was considered, linearisation was best done on a large computer because, once data has been read into this machine from a magnetic tape, the time for computing is small particularly if an array retrieval process is used.

Such a process of array retrieval is not very efficient if it is done using normal Fortran procedures. At Queen Mary College a special machine code procedure has been written for this purpose.

P.BRADSHAW added: Digital processing is justifiable for conditional sampling experiments and also for experiments involving <u>a few</u> measurements of triple products or other medium-complicated quantities. The cost of the time needed to set up analogue apparatus for triple products is as much as the cost of a few computer runs.

He and his colleagues will probably go over to analogue linearizing - a million " * * 2.22" exponentiations take a long time. But, Cheesewright's array - look up technique is also quite time-consuming although it seems the best buy for really large numbers of data points.

II. 4-5 Questions addressed to F. DURST and W. RODI Imperial College, London

M. HOFFMEISTER asks: In your new method, you have the influence of third and higher order velocity correlations on the mean velocity vector and hence on the Reynolds stresses. How large is this influence in comparison with that of conventional methods? You only pointed out the differences between two approximate methods and it is therefore not possible to know, how larger are the errors that arise.

F.DURST and W.RODI reply: For both the new and the conventional method, it is difficult to assess <u>quantitatively</u> the error introduced by the neglect of third and higher order velocity correlations; for the true values of the quantities which we seek to measure are not known, and, in general, the neglected correlations are not known with certainty either. The error will depend on the flow situation; We expect it to increase with increasing turbulence intensity. A qualitative assessment reveals that, in highly turbulent flows, the new method should introduce a smaller error than the conventional method: the third and higher order correlations are considered small compared with the mean velocity whereas the conventional method assumes they are small compared with second order correlations (Reynolds stresses).

II. 4-6 Questions addressed to H.H.BRUUN and P.A.O.L.DAVIES, Institute of Sound and Vibration Research, Southampton University

JOAN WHAMOND asks: Don't you consider that for most work rather than use a complex algebraic approximation or a tabulated array and digital techniques it would be just as satisfactory to carry out conversion of voltage to flow velocity by use of an analog function generator?

M.HOFFMEISTER asks: Why do you think that it is possible to establish a non-linear analysis of single hot-wire measurements without taking into account the fluctuating velocity component being perpendicular to the mean velocity vector and to the hot-wire? In my opinion this component has the main non-linear influence on the voltage of a hot-wire anemometer. I reported on this problem at Euromech-24 one year ago.

H.H.BRUUN and P.A.O.L. DAVIES reply to JOAN WHAMOND, both linear and non-linear analysis is influenced by higher order velocity terms. As demonstrated in Ref.(2) of paper II. 4-6 no improvement in accuracy can be obtained by linearization at low and medium intensity turbulence. Furthermore, any curve fitting by an analog function generator cannot improve the accuracy of the direct calibration data.

However, in many practical applications, time and convenience are often the dominating factors, while accuracy may be desirable but not of primary importance. A combination of analog evaluation techniques and linearization is therefore often justified in practical applications. A summary of several linearization methods, their relative acccuracy and calibration procedure is given in ref. (2) of paper II. 4-6.

To M.HOFFMEISTER, both a non-linear and a linear analysis will be influenced by the fluctuating velocity component perpendicular to the mean velocity vector and to the hot wire, as shown in ref. (2) of the paper. A mean velocity analysis in this paper demonstrates that a non-linear evaluation will give a value which is in less error than one obtained from a linear analysis. However, from Table 1 it follows that $\overline{V}_{lin}$ and $\overline{V}_{n.lin}$ differs only by 3% at a turbulence of 30%. For mean velocity and turbulence intensity measurements at low and medium turbulence intensity the values obtained by linear and non-linear analysis will therefore be nearly equal. If third or higher order fluctuation terms are to be evaluated, then linear evaluation should be used, because of the significant difference in the values obtained from linear and non-linear analysis.

II. 4-7 Questions addressed to J. NOSKIEVIC, University of Ostrava, Czechoslovakia

P.E.NIELSEN asks: Use of the method referred to in your reference (2) required low frequencies a special dummy probe. How was your dummy probe made? The measured data gives the impression that the curves are only showing the gain from generator to output and not the frequency characteristic of the anemometer.

J.NOSKIEVIC replies: A dummy probe was not used in our measurements. We have measured only the amplitude attenuation attenuation at low frequencies of the measuring equipment consisting of the hot-film probe, the constant - temperature anemometer and the power - amplifier vibrometer.

Our experimental results about the dynamic properties of hot-film anemometers not only showed good agreement with the one-dimensional characteristics elaborated by Bellhouse and Rasmussen (published in DISA Information No.6) especially for higher frequencies (over 300 Hz) but a considerable difference at lower frequencies which is of great importance when measuring turbulence in water pipes. In the written contribution the calculation is made with the one-dimensional theory elaborated by the above authors (see curve 1, fig. 1. in Volume I, page 169, where overheating ratio :

$$r = \frac{R - R_O}{R_O} = \frac{2.89 - 2.81}{2.81} = 0.0285$$

Excess temperature :

$$\Delta T = \frac{r}{\beta} = \frac{0.0285}{2.2 \times 10^{-3}} = 13^{o}C$$

Measured probe current :

$$I = 0.14 \text{ amp}$$

Heat flux :

$$Q = \frac{R.I^2}{4.2A} = 6.28 \text{ cal/scm}^2$$

Total heat transfer coefficient :

$$\alpha_t = \frac{Q}{\Delta T} = \frac{6.28}{13} = 0.484 \text{ cal/scm}^2 \ {}^{o}C$$

According to the equation

$$\alpha = \frac{\alpha_t}{1 + \frac{1}{1 + \frac{\alpha \ell}{k}}} = 0.322 \text{ cal/scm}^2 \ {}^{o}C$$

we have

$$\frac{\ell \alpha}{k} = 0.92$$

and the final level of amplitude distortion

$$R(w)_{w \to \infty} = \frac{1 + \frac{\ell \alpha}{k}}{2 + \frac{\ell \alpha}{k}} = \frac{1.92}{2.92} = 0.658$$

The point of inflexion of the curve is at $\ell\sqrt{\frac{w}{2a}} = 2.5$, which corresponds at $\frac{\ell}{\sqrt{a}} \doteq 0.080$s to the frequency $f \doteq 310$ Hz. The experimental results are in good agreement (curves on fig.1) with this one-dimensional theory.

When comparing our experimental results with the one-dimensional theory at lower frequencies the measured amplitude distortion of magnetohydrodynamic measurements rises whilst that produced by the hot-film anemometer falls. The difference was probably caused by the measuring instruments.

Written Contribution by J.T. TURNER, University of Manchester, on the Yaw Response of Hot Wires

This comment relates to statements made in papers II. 4-1, II. 4-2 and II. 4-5 regarding the yaw response of hot-wire probes. It is especially relevant to measurements at high levels of turbulence where appreciable angular deviations of the instantaneous velocity vector from the calibration direction may occur.

Two alternative expressions can be used to describe the yaw sensitivity of the hot wire probe. For convenience, we can refer to these as the k factor expression,

$$E^2 - E_0^{\ 2} = BU^n \ (\cos^2 \theta + k^2 \sin^2 \theta)^{n/2} \tag{1}$$

and the cosine law relationship,

$$E^2 - E_0^{\ 2} = BU^n \cos^m\theta \tag{2}$$

The voltage E_0 occurs at zero velocity and the voltage E at a (steady) mean velocity U and a yaw angle θ. The parameters B, n, k, m, may be determined expirically for a given probe geometry and will normally, although not always (see references 1, 2), be taken as constant within a restricted velocity range. The intention here is to stress the need for caution in choosing the most accurate expression from equations (1) or (2). For this purpose, reference will be made to some results obtained by the writer.

Experimental values have been obtained for 5 μ-diameter tungsten (normal and slant) wire probes (1/d = 240 approximately) in a low turbulence calibration wind tunnel. The wire element itself was used to determine the normal flow direction (θ = 0) by locating the angular positions of the two voltage minima (θ = ± 90-deg. approximately) - see figures 1 and 2. In combination with the device used to rotate the probe, this gave an angular accuracy of better than 0.2 degree at each yaw setting.

Noting that flow normal to the wire gives

$$E_N^{\ 2} - E_0^{\ 2} = BU^n \tag{3}$$

it is possible to obtain an explicit form for the parameters of interest.

$$\text{Thus} \quad K = \sqrt{\frac{A - \cos^2\theta}{\sin^2\theta}} \quad \text{with } A = \left(\frac{E^2 - E_0^{\ 2}}{E_N^{\ 2} - E_0^{\ 2}}\right)^{2/n} \tag{4}$$

$$\text{and} \quad m = \log\left(\frac{E^2 - E_0^{\ 2}}{E_N^{\ 2} - E_0^{\ 2}}\right) (\log \cos \theta)$$

FIGURE 1. YAW RESPONSE OF A NORMAL HOT WIRE PROBE.

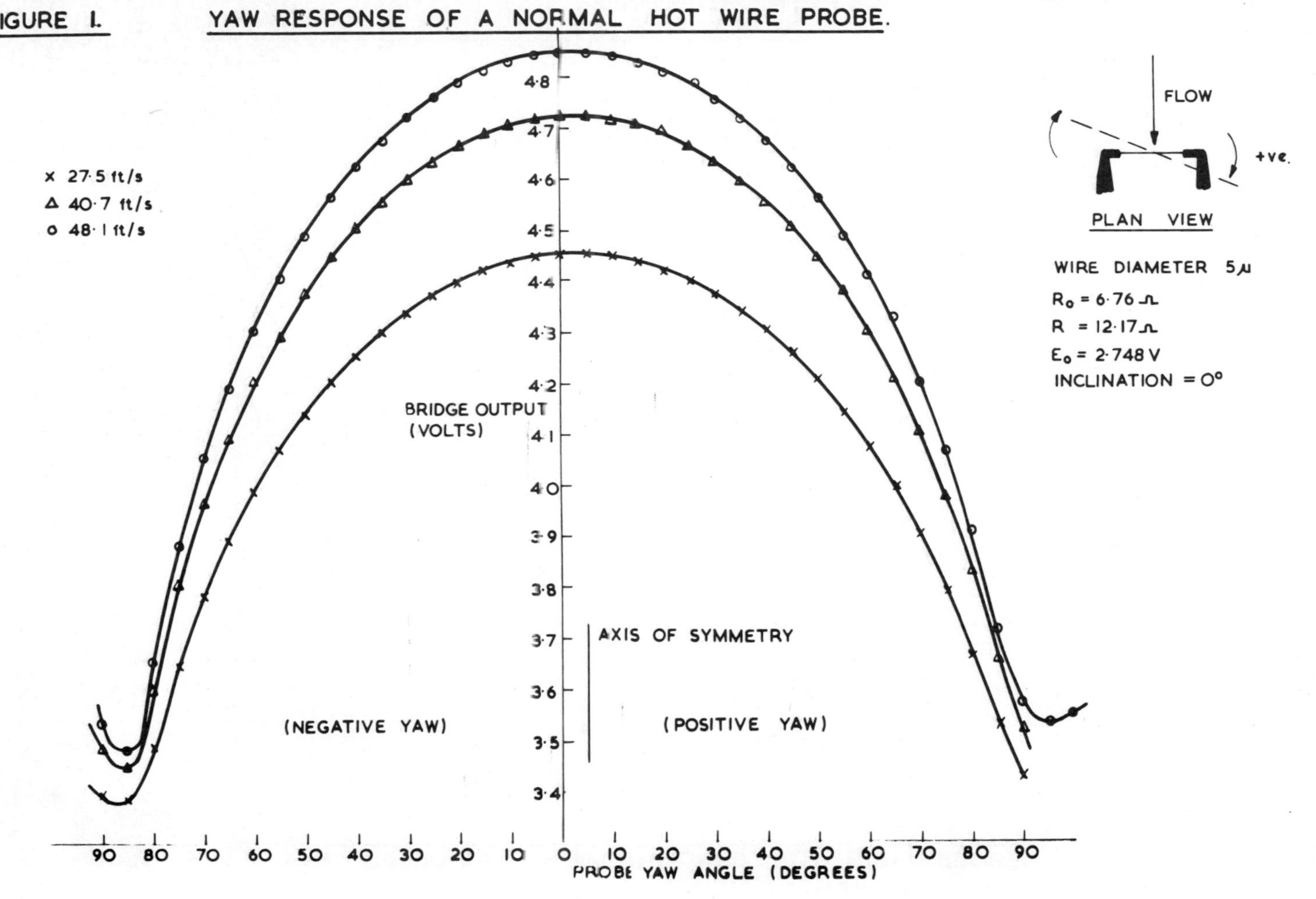

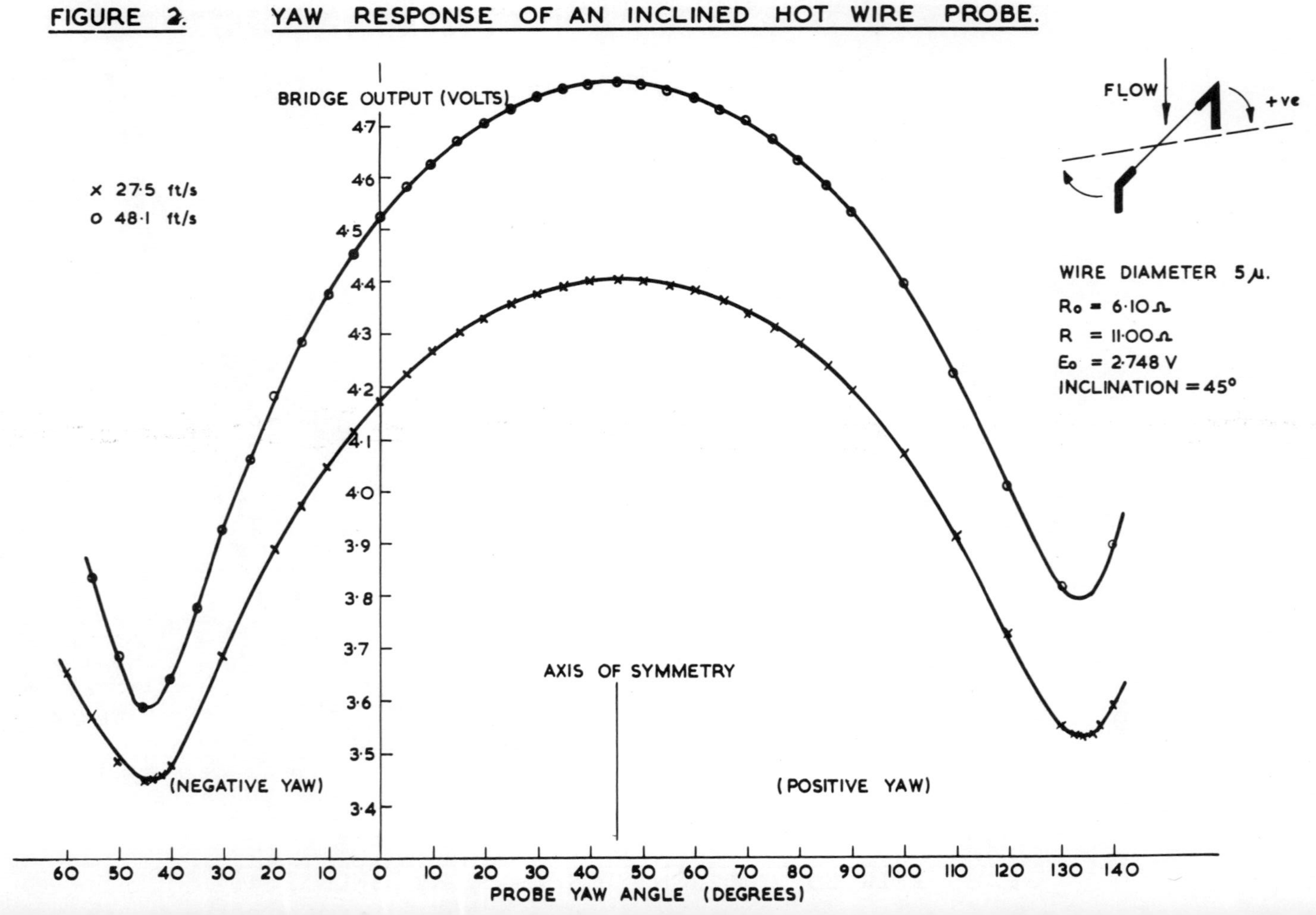

FIGURE 2. YAW RESPONSE OF AN INCLINED HOT WIRE PROBE.

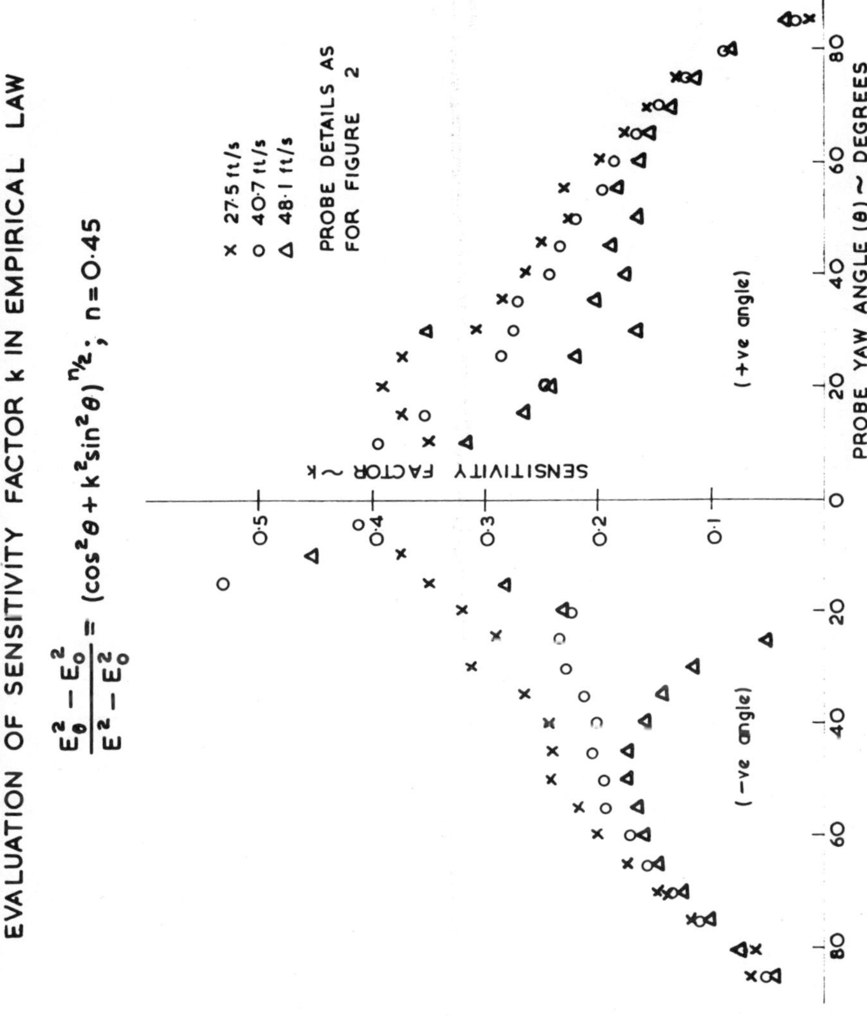
FIGURE 3
YAW RESPONSE OF A NORMAL HOT WIRE PROBE
EVALUATION OF SENSITIVITY FACTOR k IN EMPIRICAL LAW
$\frac{E_\theta^2 - E_0^2}{E^2 - E_0^2} = (\cos^2\theta + k^2\sin^2\theta)^{n/2}$; n = 0·45
× 27·5 ft/s
○ 40·7 ft/s
△ 48·1 ft/s
PROBE DETAILS AS FOR FIGURE 2
SENSITIVITY FACTOR ~ k
(+ve angle)
(−ve angle)
PROBE YAW ANGLE (θ) ~ DEGREES
0·5
0·4
0·3
0·2
0·1
0
20
40
60
80

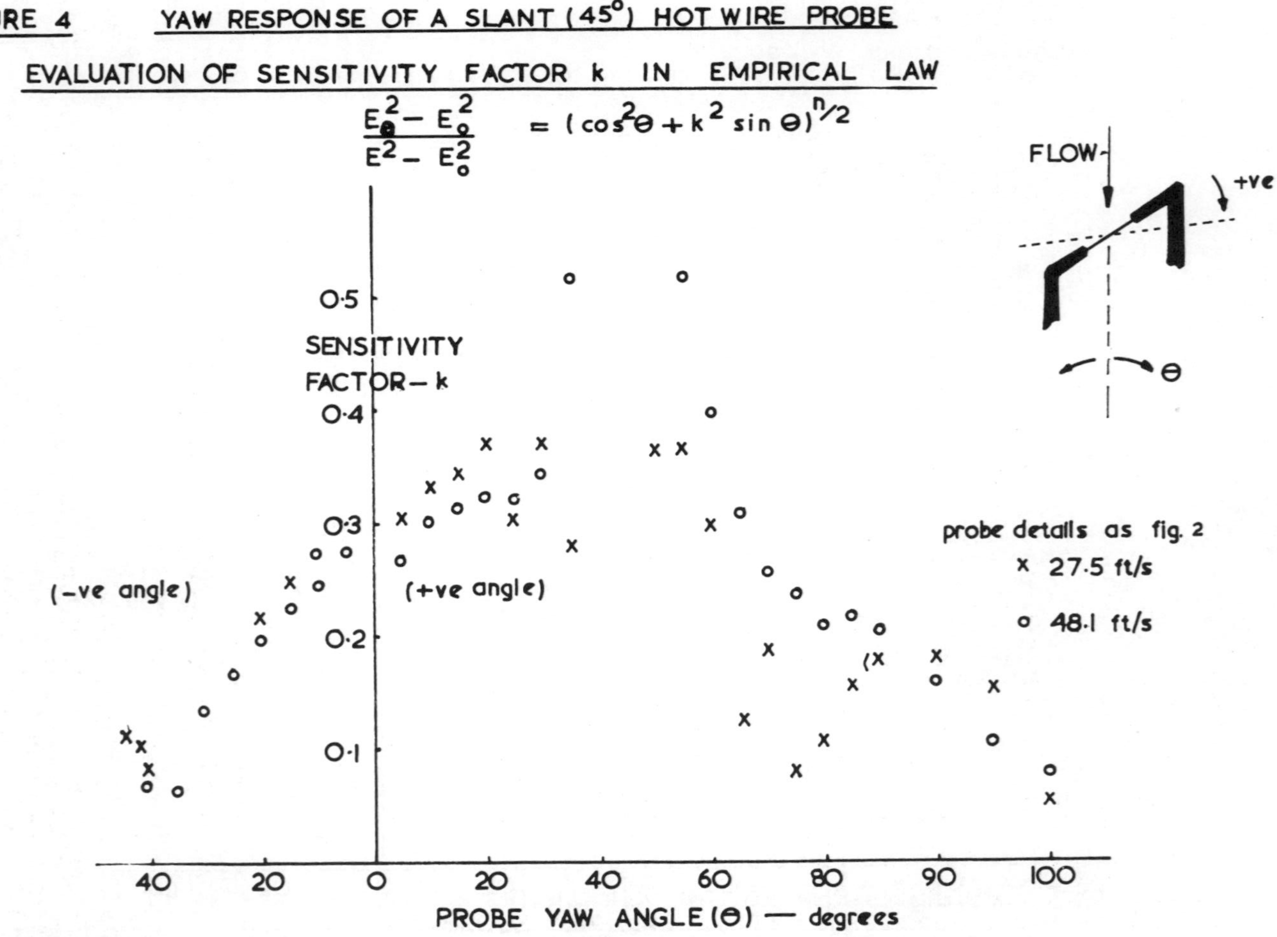

FIGURE 4 YAW RESPONSE OF A SLANT (45°) HOT WIRE PROBE

FIGURE 5 YAW RESPONSE OF A NORMAL HOT WIRE PROBE

IMPORTANCE OF ASSUMED EXPONENT VALUE n IN CALCULATED k FACTOR

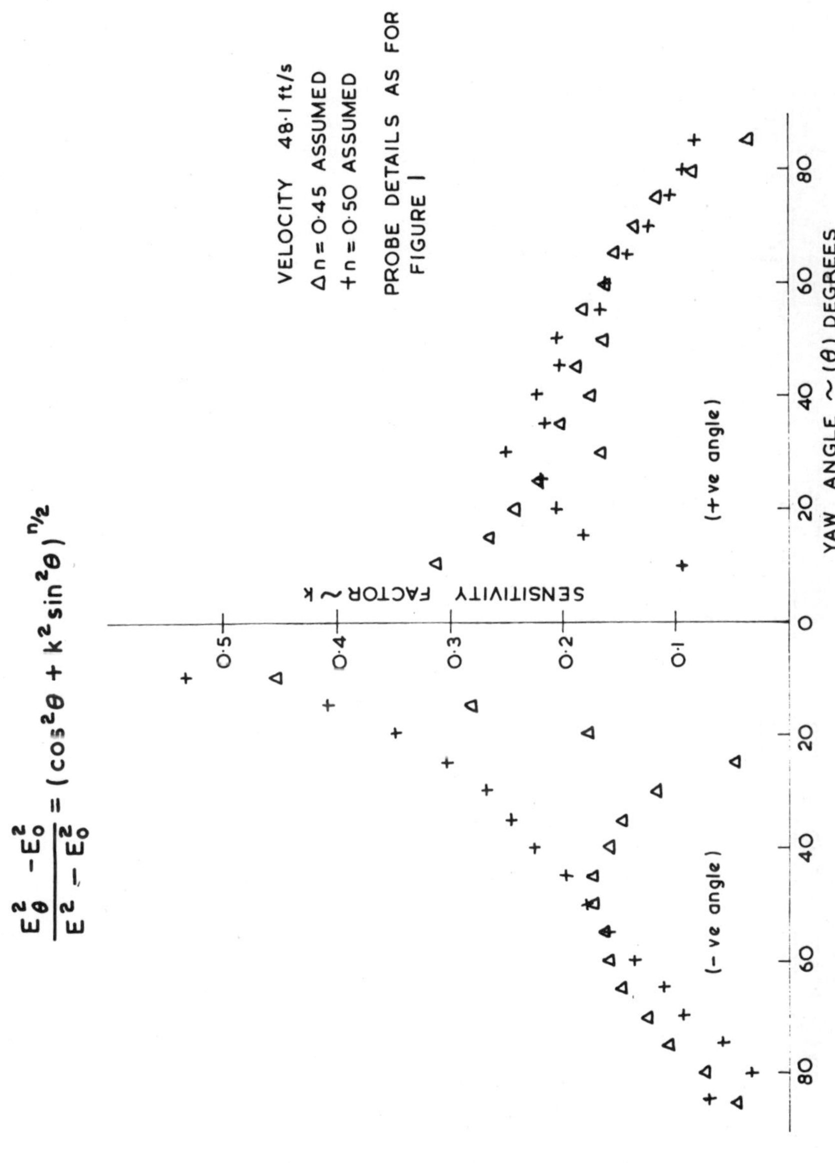

FIGURE 6 YAW RESPONSE OF A NORMAL HOT WIRE PROBE

EVALUATION OF EXPONENT (m) IN THE EMPIRICAL LAW $\frac{E_\theta^2 - E_0^2}{KU^n} = \frac{E_\theta^2 - E_0^2}{E^2 - E_0^2} = \cos^m\theta$

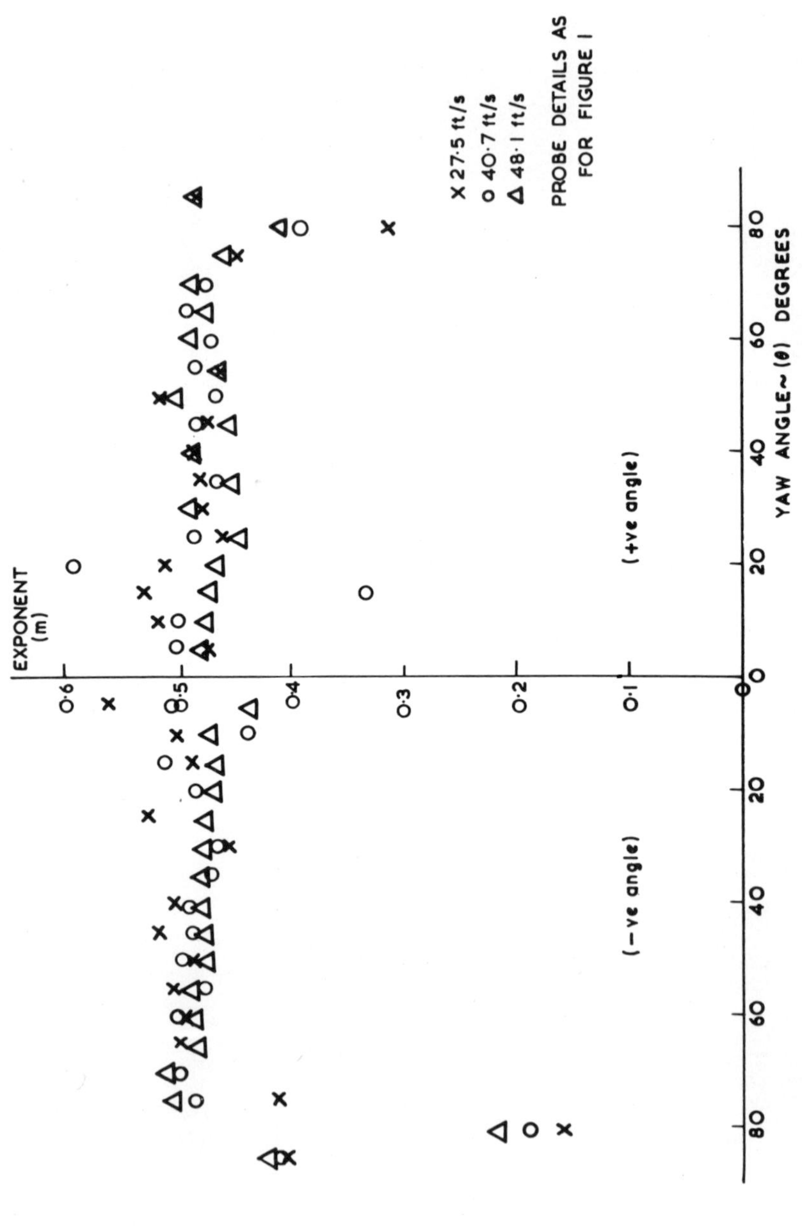

EXPONENT (m)
0·6
0·5
0·4
0·3
0·2
0·1
0
(+ve angle)
(−ve angle)
YAW ANGLE ~ (θ) DEGREES
80 60 40 20 0 20 40 60 80
X 27·5 ft/s
o 40·7 ft/s
Δ 48·1 ft/s
PROBE DETAILS AS FOR FIGURE 1

FIGURE 7 YAW RESPONSE OF A SLANT (45°) HOT WIRE PROBE.

EVALUATION OF EXPONENT (m) IN THE EMPIRICAL LAW $\frac{E_\theta^2 - E_o^2}{KU^n} = \frac{E_\theta^2 - E_o^2}{E^2 - E_o^2} = \cos^m\theta$

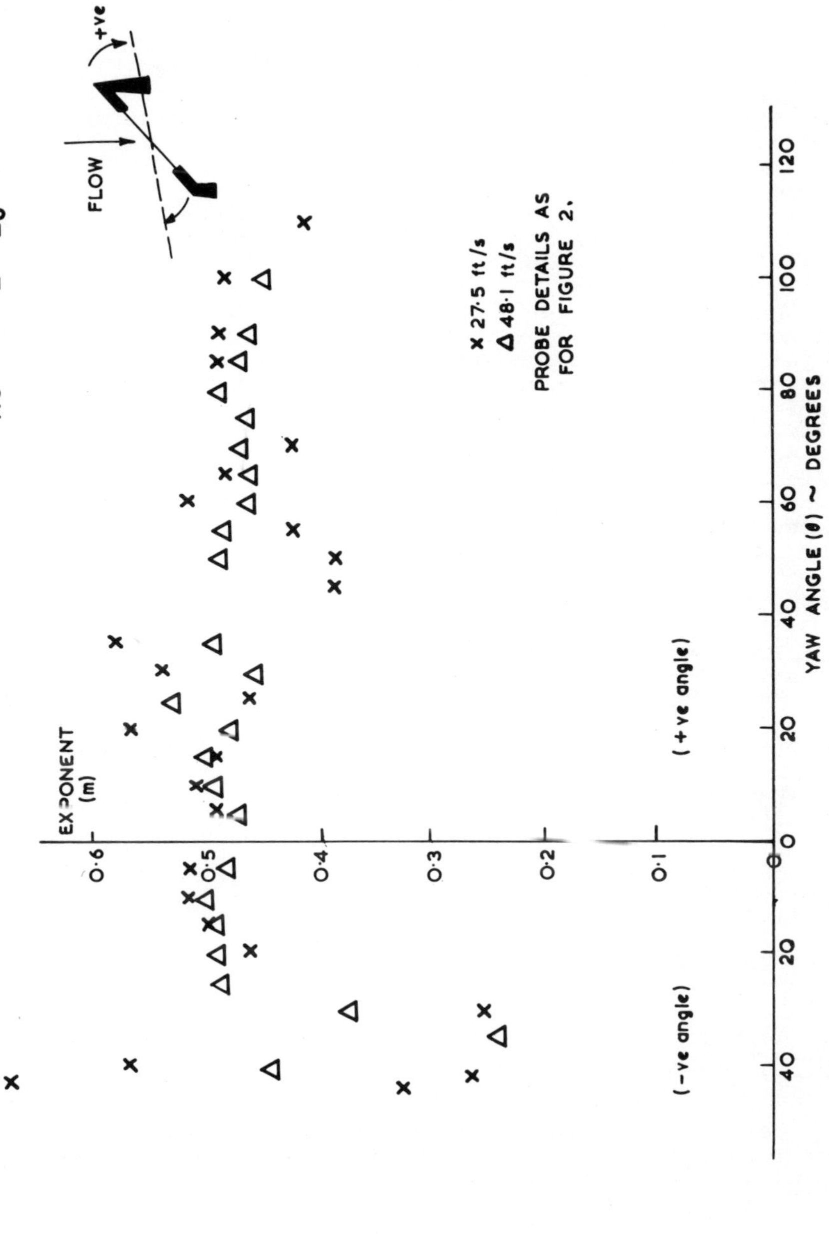

The expression for k depends on the velocity exponent n, whereas m is seen to be independent of the value assumed. This is a point in favour of the cosine law.

Analysis of figures 1 and 2 leads to the results shown in figures 3, 4, 5, 6 and 7. These show that k is not even constant for the individual probes but varies significantly with yaw angle and the flow velocity (or assumed exponent n). This behaviour should be contrasted with the measured values for the exponent m which remains constant, within the experimental scatter, to yaw angles greater than 70-deg. These results confirm the findings of Davies and Bruun (4) and Bruun (5). The variations in k are also referred to by Jørgensen (6).

Despite an earlier flirtation with expression (1) (see Turner (7)), the writer is now convinced of the desirability of using the cosine law (equation 2) in preference to the alternative approximation (1) in hot-wire analysis. Moreover, the use of digital techniques described by Cheesewright (3) and others at this Conference could be enormously simplified if empirical laws of suitable accuracy were available. Any attempt to describe the effect on hot-wire response of the higher order terms must incorporate accurate yaw response relationships, otherwise it seems certain that additional and very significant errors may be introduced.

References

1. Paper II. 4-1 by Dvorak and Syred, Volume I.

2. Paper II. 4-2 by Cheesewright, Volume I.

3. Paper II. 4-5 by Durst and Rodi, Volume I.

4. Davies and Bruun. The performance of a yawed hot wire. Symposium at N.P.L. November 1968.

5. Bruun. Interpretation of a hot-wire signal using a universal calibration law. J.Sci.Inst.,March 1971; also Paper II. 4-6, Volume I.

6. Jørgensen, F.E. Directional sensitivity of wire and fibre probes. Disa Information. No. 11, May 1971.

7. Turner, J.T. Non-linearity corrections for the response of a hot wire anemometer at high levels of turbulence. Symposium at N.P.L., November 1968.

Comment on this written contribution by H.H.BRUUN. J.T.Turner has presented calibration data for a yawed hot-wire probe which shows a strong yaw dependence of k in his equation (1). However m in equation (2) was found to be independent of the yaw angle θ.

A recent investigation (to be published) of the effect of these variations shows that for practical applications the difference in results by these two methods is negligible. Writing the effective cooling velocity V_e as

$$V_e = U f(\alpha)$$

where $f(\alpha) = (\cos^2 \alpha + k^2 \sin^2 \alpha)^{\frac{1}{2}} = \cos^m \alpha$ gives the first order corrections to both mean and fluctuating velocities.

When k and m were evaluated in the investigation the additional corrections for fluctuation quantities were found to be

$$A = \frac{1 - k^2}{1 + k^2} \quad \text{and } m$$

For 1 mm, 5-μ tungsten hot wires with a parallel support the difference in value between A and m was found to be 1%. For turbulence measurements therefore, the total effect of the uncertainty in yaw is only of the order of 1%.

II. 4-8 Questions addressed to J.R.THOMAS, D.R.BROWN and A.D.BIRCH, The Gas Council, Solihull

P.A.O.L.DAVIES asks: Would you comment further on the use of a constant value for k in the response equations (1) to (4)?

P.BRADSHAW asks: Why did you assume a square wave form for the fluctuating component of the wire voltage, E' ?

R.CHEESEWRIGHT asks: Had the authors of this paper plotted their turbulent fluctuation data normalised against the local velocity rather than a mean velocity and thus realised the very high levels of turbulence which they had apparently measured?

D.DURAO asks: There are investigators saying that it is impossible (or at least very difficult) to separate the temperature effect when measuring velocities in high temperature flows. In your paper you refer to the turbulence intensity problem but not to the temperature one. Can you give more information how did you treat this problem?

J.R.THOMAS, D.R.BROWN and A.D.BIRCH reply: to P.A.O.L.DAVIES, we have previously examined the wire response to a yawed velocity and a typical result is shown in Fig 1 (θ is defined as the angle between the normal to the wire and the incident velocity vector.) Figure 2 shows the variation in k with velocity for two types of probe. Although the variation in k is relatively large the effect on the velocity profile can be shown to be very small. This is illustrated in Fig. 4 which compares the mean axial flow vector profile obtained using two different values of k. (Um is the inlet central line velocity). Further evidence of the reliability of the method for obtaining mean velocity information in turbulent flows can be seen from Fig. 3, where the distribution in mean axial velocity derived from the hot wire is compared with that produced from a laser fringe anemometer. It is seen that the agreement is within a few per cent.

To P.BRADSHAW, the validity of the square wave form assumption has previously been examined by Davies*. He has shown that other assumed waveforms (incorporating a sawtooth and sine wave) lead to substantially similar values of turbulence intensity up to around 60%. It would therefore seem that the particular form assumed for the shape of E' is rather unimportant. However, we are examining the effect of other equations describing E' at present.

* T.W.Davies: A study of the Aerodynamics of the Recirculation Zone Formed in a Free Annular Air Jet. Ph.D. Thesis, University of Sheffield, 1969.

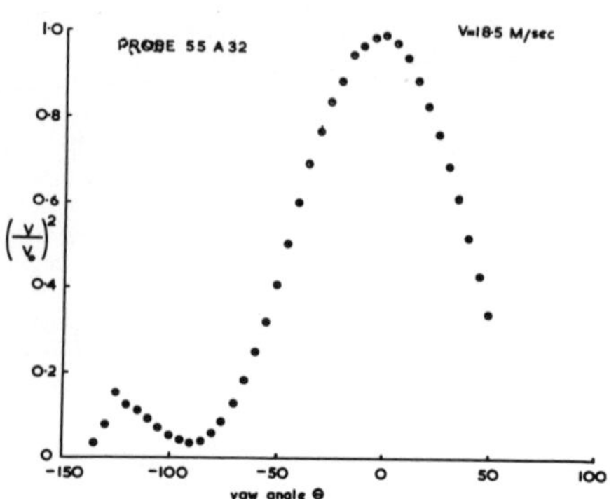

FIG. 1. EFFECT OF YAW ON VELOCITY

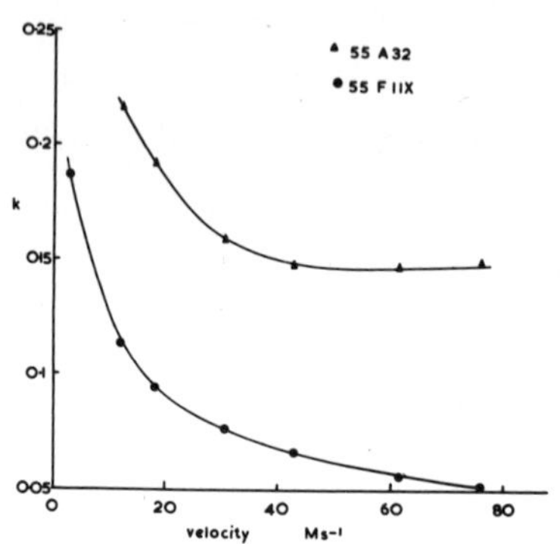

FIG. 2. VARIATION OF PARAMETER K WITH VELOCITY

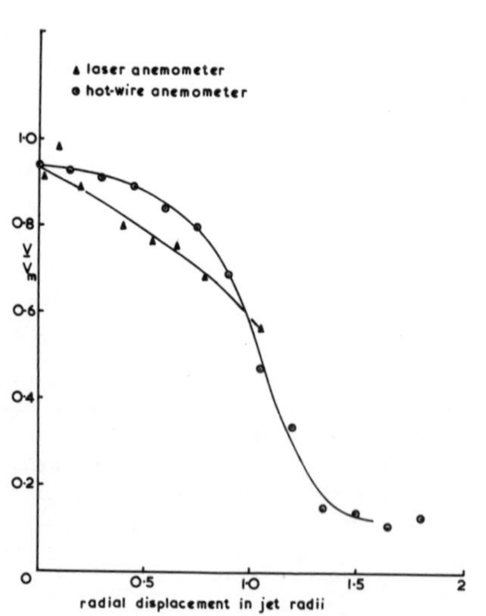

FIG. 3. MEAN VELOCITY PROFILES IN A SUDDEN ENLARGEMENT

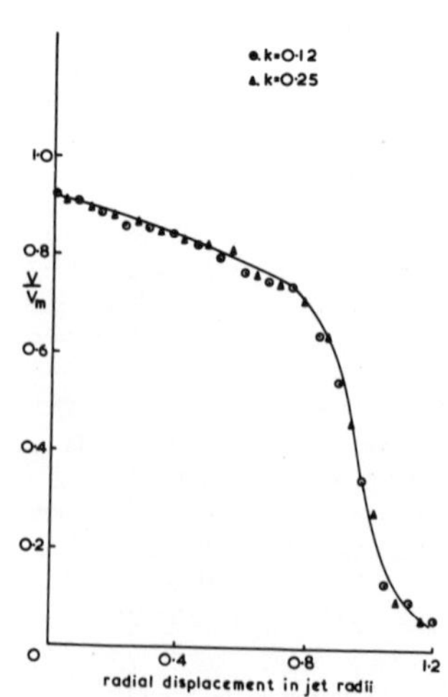

FIG. 4. EFFECT OF K ON MEAN VELOCITY

To R.CHEESEWRIGHT, although we are principally interested in the distribution of the kinetic energy of turbulence we derive both $\sqrt{U'^2}/\bar{U}_{mean}$ and $\sqrt{U'^2}/\bar{U}_{local}$ in our analysis. The results from the models discussed in our paper indicate that the maximum $\sqrt{U'^2}/\bar{U}_{local}$ is around 55% in regions close to the wall of the tunnel. Here the contribution made by the rms velocity (as can be seen from fig. 3 of the paper) to the integrated turbulent kinetic energy across a plane at right angles to the flow is relatively low.

To D.DURAO, we were not concerned with thermal effects in the paper which reports the results of aerodynamic measurements in isothermal flow, although we have done some work in high temperature systems using a laser fringe anemometer. In such a case we would expect that the frequency at which particles pass through regions of light and dark in the set of fringes would be considerably greater than the frequency of the local temperature fluctuations.

A general written comment on mean and fluctuating velocity measurements by hot-wire and hot-film techniques from P.O.A.L. DAVIES.

We are concerned here with measurements in industrial environments and although I admire the persistence and the skill of the authors of many of the papers in this session I wish to amplify P. Bradshaw's comment that such complexity is often unjustified.

We choose simple laws for practical reasons and these are adequate over reasonably wide velocity ranges so that analogue linearisations are satisfactory for most practical purposes. Dynamic calibration is seldom really necessary so the problems raised by Perry and Morrison (see for example II.4-1) do not arise when calibration and experiment are performed under the same conditions.

To use digital conversion commercially software must be written for the computer and this is seldom justified. However, in cases such as for low flow velocities, where simple analytic approximations to the cooling law are difficult to find the use of techniques like that of R. Cheesewright becomes much more attractice.

If arrays are employed great care must be taken with the interpretation of results since many of the constants used in fitting the calibration data are insufficiently defined to be used with confidence. Consider, for example, J. Turner's demonstration that k varies with flow speed and yaw angle. Unfortunately Champagne and his colleagues made their comprehensive measurements on many different probe types at one flow speed and a limited range of yaw angle. Since results critically depend on the values of such constants, the interpretation of such measurements is open to serious question.

II.5 OTHER TECHNIQUES FOR MEAN AND FLUCTUATING VELOCITY MEASUREMENTS

Further written comment on the paper, magnetohydrodynamic measurement of velocity fluction, by J. JANALIK.

Grossman (ref. 2 in paper II.5) compared the results of relative intensity of turbulence obtained by means of the magneto-hydrodynamic method with those achieved by Laufer (ref. 3 in paper

II.5) using a constant temperature anemometer in air. The lesser magnitudes obtained by the first method Grossman explains as follows:

The equation

$$\text{grad}\ \phi = v \times B - \frac{i}{\gamma} \qquad (5)$$

Factors into components

$$\frac{\partial\phi}{\partial x} = B.V - \frac{i_x}{\gamma}$$

$$\frac{\partial\phi}{\partial y} = - B.U - \frac{i_y}{\gamma}$$

$$\frac{\partial\phi}{\partial z} = - \frac{i_z}{\gamma} \qquad (6)$$

Assuming that

$$v = \overline{v} + v' \ , \ i = \overline{i} + i' \qquad (7)$$

and applying the Reynolds rates of averaging we have

$$\overline{\left(\frac{\partial\phi}{\partial x}\right)^2} = B^2\left(\overline{V^2} + \overline{v^2}\right) - \frac{2B}{\gamma}\left(\overline{Vi_x} + \overline{vi'_x}\right) + \frac{\overline{i_x}^2 + \overline{(i'_x)^2}}{\gamma^2}$$

$$\overline{\left(\frac{\partial\phi}{\partial y}\right)^2} = B^2\left(\overline{U^2} + \overline{u^2}\right) + \frac{2B}{\gamma}\left(\overline{Ui_y} + \overline{ui'_y}\right) + \frac{\overline{i_y}^2 + \overline{(i'_y)^2}}{\gamma^2}$$

$$\overline{\left(\frac{\partial\phi}{\partial z}\right)^2} = \frac{\overline{i_z}^2 + \overline{(i'_z)^2}}{\gamma^2} \qquad (8)$$

The expressions for the mean square of the fluctuating potential gradients are then

$$\overline{\left(\frac{\partial\phi'}{\partial x}\right)^2} = B^2.\overline{v^2} + \frac{\overline{(i'_x)^2}}{\gamma^2} - \frac{2B}{\gamma}\ \overline{v.i'_x}$$

$$\overline{\left(\frac{\partial\phi'}{\partial y}\right)^2} = B^2.\overline{u^2} + \frac{\overline{(i'_y)^2}}{\gamma^2} + \frac{2B}{\gamma}\ \overline{u\,i'_y} \qquad (9)$$

$$\overline{\left(\frac{\partial\phi'}{\partial z}\right)^2} = \frac{\overline{(i'_z)^2}}{\gamma^2}$$

Grossman explainsthe differences in the relative intensity of turbulence velocity u by neglecting the term in equation (9)

$$\frac{\overline{(i_y')^2}}{\gamma^2} + \frac{2B}{\gamma} \quad \overline{ui_y'}$$

the author has observed that

$$\frac{\overline{(i_y')^2}}{\gamma^2}$$

can be considerably reduced if an amplifier with a great internal resistence is used.

The influence of the correlation moment $\frac{2B}{\gamma}$ $\overline{ui_y'}$ is difficult to estimate and can be one reason why the magnetohydrodynamic method gives a lower evaluation. One factor which diminishes the induced voltage according to Shercliff (ref. 1 in II.5) is the so called "end effect" (boundary limit or boundary condition), for the measuring is not executed in a magnetic field of infinite space as suggested by the equation but in a relatively smaller space.

III Measurement of voidage

III.1 Questions addressed to D.McWILLIAM, Dounreay Experimental Reactor Establishment and T.F. ROYLANCE, University of Nottingham

J.M.DELHAYE asks: 1. How do you choose your threshold level?

2. Did you check your results by a global measurement and by the rays absorption method?

D.McWILLIAM and T.F. ROYLANCE reply: 1. The effect of threshold level on indicated void fraction and bubble frequency is shown in fig. (7) on P.190 of Vol I of the Proceedings and in the attached figure, fig. (8). It is apparent that if the threshold voltage is reduced below a point denoted by A in fig. (7) the effect of background noise causes the indicated void fraction to rise very steeply. This is mentioned on page 186. Fig. (8) shows the relationship between indicated void fraction and bubble frequency (with threshold level decreasing to the right). It can be seen that too low a reference level results in a sharply increased degree of scatter.

2. The results were checked by carrying out three traverses of voidage and velocity at 120-deg. spacing in a flow which was simultaneously monitored using the average void fraction meter described on pages 183-184. Integration of the local profiles and comparison with the average values showed the local meter to read some 25-60% low depending upon the bubble size. Typical results are shown in fig. (9).

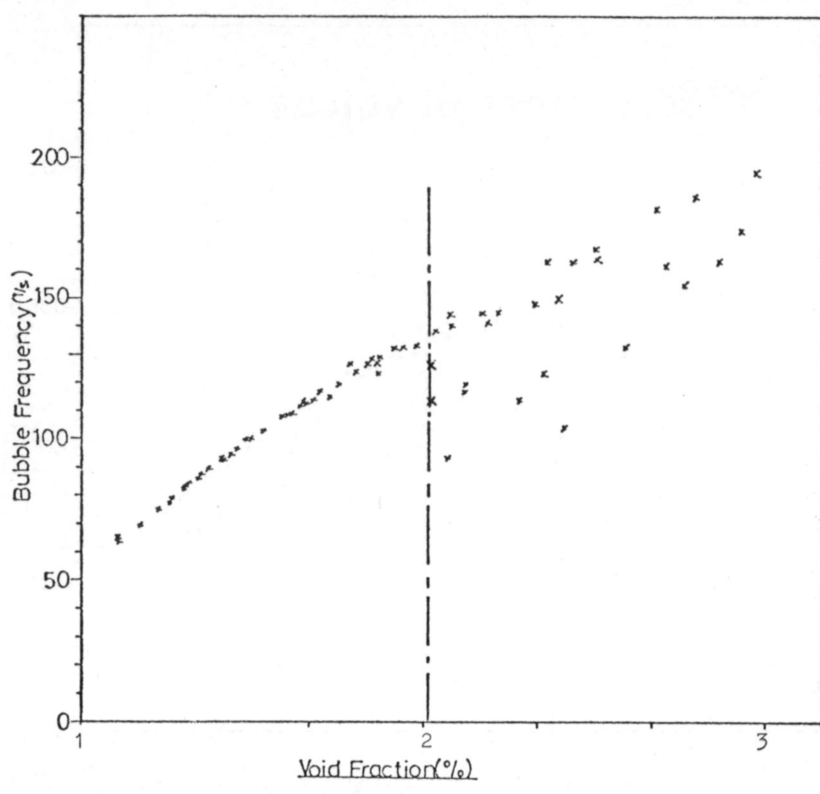
Bubble Frequency (1/s)
200
150
100
50
0
1
2
3
Void Fraction(%)

FIG. 8.

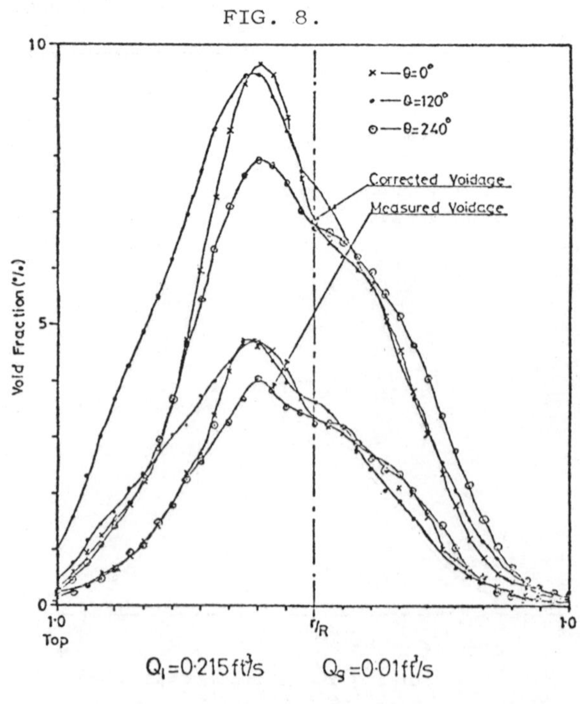
10
5
0
Void Fraction (%)
× — θ=0°
• — θ=120°
o — θ=240°
Corrected Voidage
Measured Voidage
1·0
Top
r/R
1·0
Q_l=0·215ft³/s
Q_g=0·01ft³/s

FIG. 9.

Written contribution by H. LECROART.

In our laboratory at Saclay we developed and we use the miniature probe shown in fig. 4 of J.M. Delhaye's paper, III.2. We also developed the electronics associated with this "variable impedance" (as distinct from variable resistance) probe. Our circuit seems to us to give decisive advantages over that used by D. McWilliam and T.F. Roylance, III.1. These are associated with time constants, shape of the analogue signal and minimum time error in transforming the analogue signals into logic signals. What produces these differences is that the feed-back impedance of the circuit is constant whilst the input variable impedance is the probe. So the probe works with constant voltage and the effects of the parasitic and fixed capacitances are cancelled. By this method we can achieve rise times of 0.5 s, of great consequence if measurements are to be made at high velocity. The work is fully reported in:

H.Lecroart and R. Porte, Electrical probes for study of two-phase flow at high velocity, Paper 3-11, International Symposium on two-phase systems, Technion, Israel, 1971.

IV Stress determination

IV - 3. Further written contribution from F.B. HOWARD and M.T. THEW

In addition to the vaned disc for which results were shown on fig. 3 in paper IV.3, we have also tested slotted discs. One of the slotted discs is that shown on the left of fig. 6 and the other was similar though with 8 wider slots. The depth of the slots were 0.10 inch and 0.20 respectively.

The 16-slot disc gave results which were to some extent a mirror image of those for the vaned disc (which is also shown in fig. 6). With the nitrogen blowing technique a marked pressure peak was shown on the slot trailing face, but this was largely attenuated before reaching the piezo-electric transducer in the wall. Other evidence has demonstrated a high radial outflow velocity in the slot, and a Coriolis effect has been estimated to account for most of the pressure rise. As the interface moves outwards, corresponding to pressure traces at the right of the figure, the pressure spike is only slightly reduced. This is in contrast to the vaned disc where the trough in pressure had its amplitude halved.

On figure 5 there is somewhat more agreement between the two techniques for the wider slot, though the wall results appear to have a phase lag. The cause of the pressure spike half way across the slot which is shown only at the wall, is postulated as a shear-flow driven vortex in the rear of slot with its axis along the slot.

The ring interface which has been referred to earlier in the paper is illustrated on fig. 7, where there is air inside the ring. This was taken with a single 0.5J argon shielded spark source with a nominal time of 1 μs. What might appear to be air bubbles inside the interface are water drops adhering the the perspex front housing.

IV - 5. Questions addressed to P.A.O.L. DAVIES and G.R. KIMBER, Institute of Sound and Vibration Research, Southampton University

D.L.SCHULTZ asks: I must emphasise that the theory of Bellhouse and Schultz referred to by Mr. Kimber is a simple one-dimensional theory which indicates the nature of the very low and high frequency behaviour of thin film skin friction gauges. It is a little early yet to claim complete understanding of the phenomenon but would you not agree that it is encouraging that the agreement with experiment is as good as it is?

T.B.MORROW asks: How was the shear stress determined during calibration of the heated thin film gauge and what was the

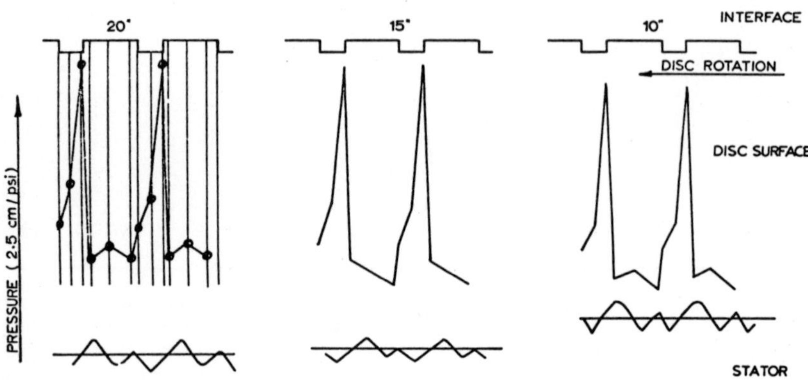

COMPILATION OF NITROGEN BLOWING RESULTS - 16 SLOT - 1/4" WIDE FIGURE. 4

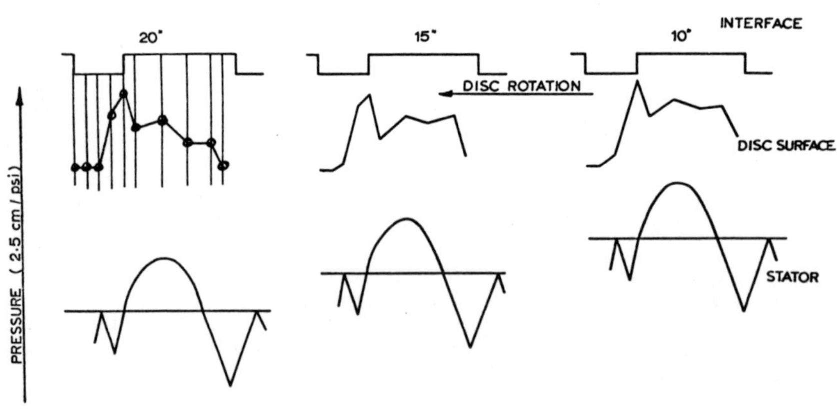

COMPILATION OF NITROGEN BLOWING RESULTS - 8 SLOT - 1/2" WIDE FIGURE 5

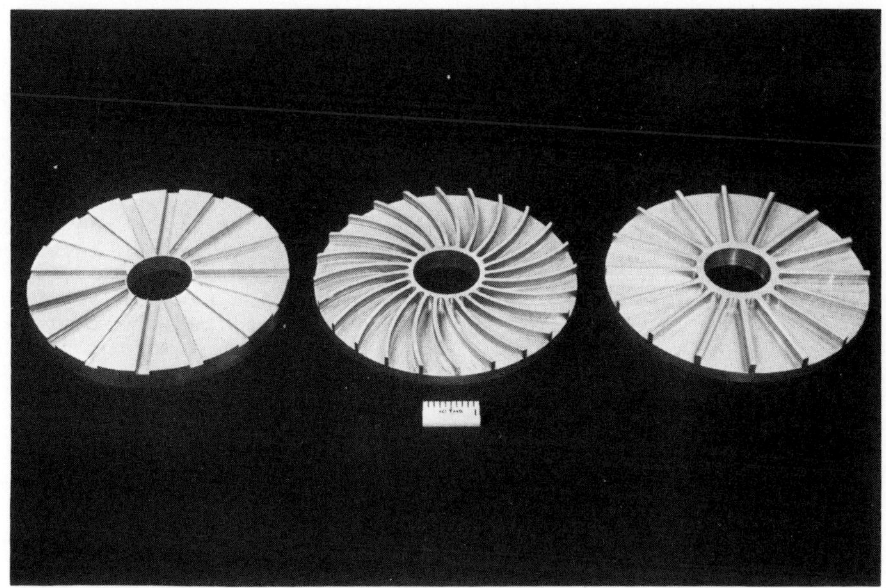

FIG.6. SEALING DISC GEOMETRY

FIG.7. WATER/AIR INTERFACE IN SEAL AT 3000 REV/MIN.

experimental uncertainty?

J.T.TURNER asks: Would the authors care to comment on the use of surface Pitot tubes to measure mean shear stress? These would seem to be both useful for calibration purposes and sufficiently rugged for use in industrial situations.

P.A.O.L.DAVIES and G.R.KIMBER reply: To D.L.SCHULTZ, we are extending our experimental work to test the theory over a wider range. Taken as a limiting case of an infinite substrate the Bellhouse and Schultz theory is in agreement with ours, as best we can judge.

To T.B.MORROW, because a large range of shear stress was covered in the experiments three methods were used to determine the stress accurately. For low shear stresses the laminar boundary layer on a flat plate was used and the stress determined from calculation A turbulent boundary layer in the same plate was used to extend the range of shear stress further. In this case the shear stress was measured by a Preston tube. For higher shear a narrow channel (80 mm x 5 mm) was used and the stress determined from the static pressure drop across the film. The air flow in the channel was turbulent and the velocity profile was fully developed. The experimental uncertainty was estimated to be within 5% in all cases.

To J.T.TURNER, we appreciate that surface tubes do have advantages over films where steady flow measurements are being made and these were readily apparent as we made use of some of these methods during our experiments. There are occasions when, even in steady flow, films have obvious advantages, particularly if it is important that the flow is not disturbed or if access to the surface is limited.

The following further written contribution was made by P.O.A.L. DAVIES and G.R.KIMBER.

When paper IV-5 was written we had made only a limited number of measurements. We now present, in figures 1 and 2, longitudinal and transverse temperature distributions which are in general agreement with recent measurements by Pope and which indicate the extent of the gauge's thermal footprint. A definite asymmetry can be observed in the profiles caused by the flow. This also appears in the measurements. The error introduced by our first approximation can be seen in figure 3, which shows the degree of asymmetry in the temperature profiles. The flow is from left to right. Figures 4 and 5 show the agreement between prediction and measurement found for a number of gauges, only the first approximation being considered. Although there must be a finite output at zero flow, figure 5 shows how closely theory matches measurement at the lowest velocity at which meaningful measurements could be made. Because of local convection cells, this was not possible at zero flow.

Turning now to dynamic behaviour, some frequency response measurements for harmonic perturbation at the amplifier input, in which a flat response for fluctuating shear stress is represented by a rise at 6dB per octave, are compared in figure 6 with measurements for a hot wire, a wire lying on the surface and an embedded wire. Results accord with those by Bellhouse and Schultz and which were predicted by their one-dimensional theory. This feature of the response is caused by the thermal inertia of the quite large substrate. Theoretical predictions are compared with measurement in figure 7.

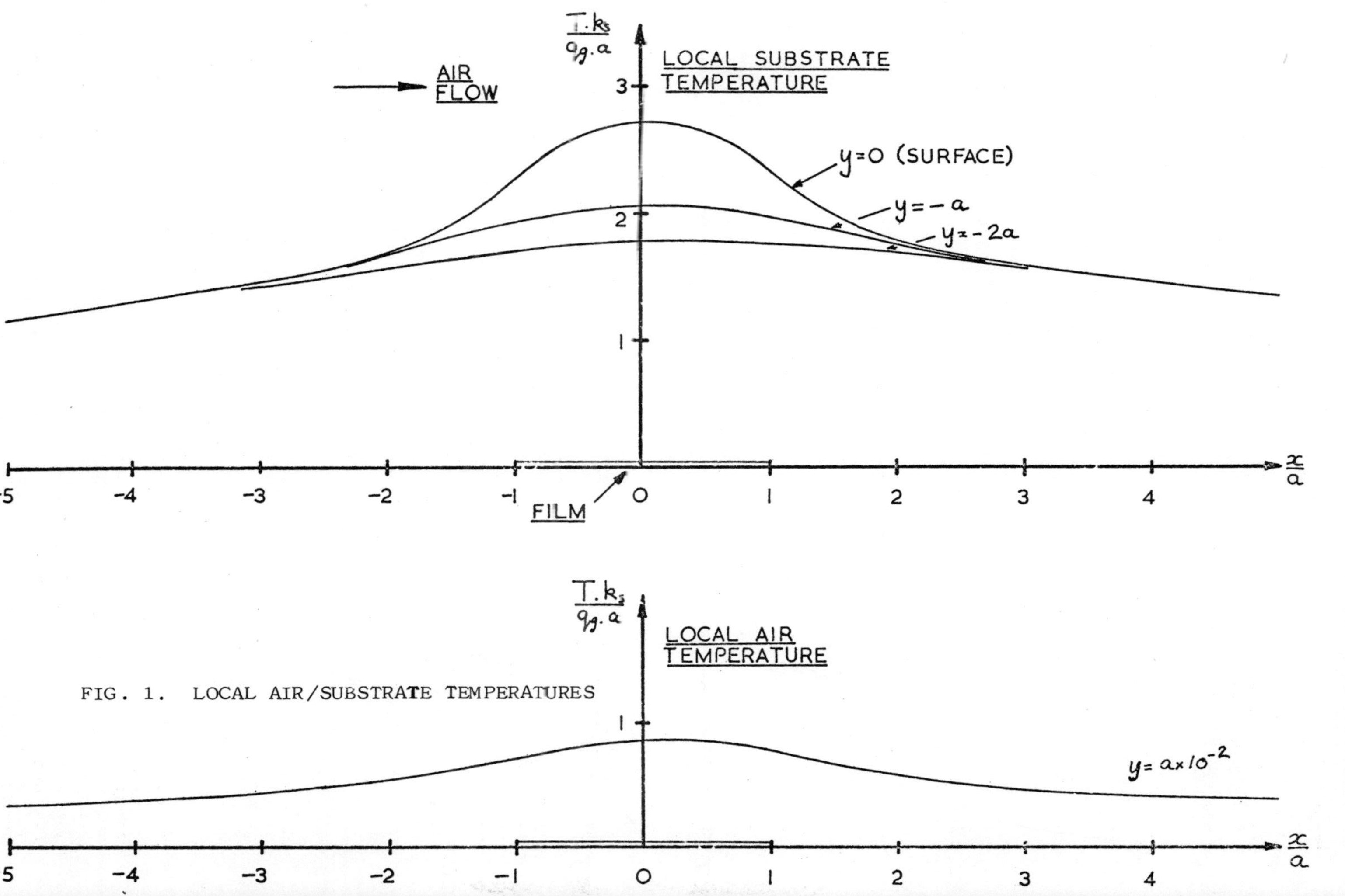

FIG. 1. LOCAL AIR/SUBSTRATE TEMPERATURES

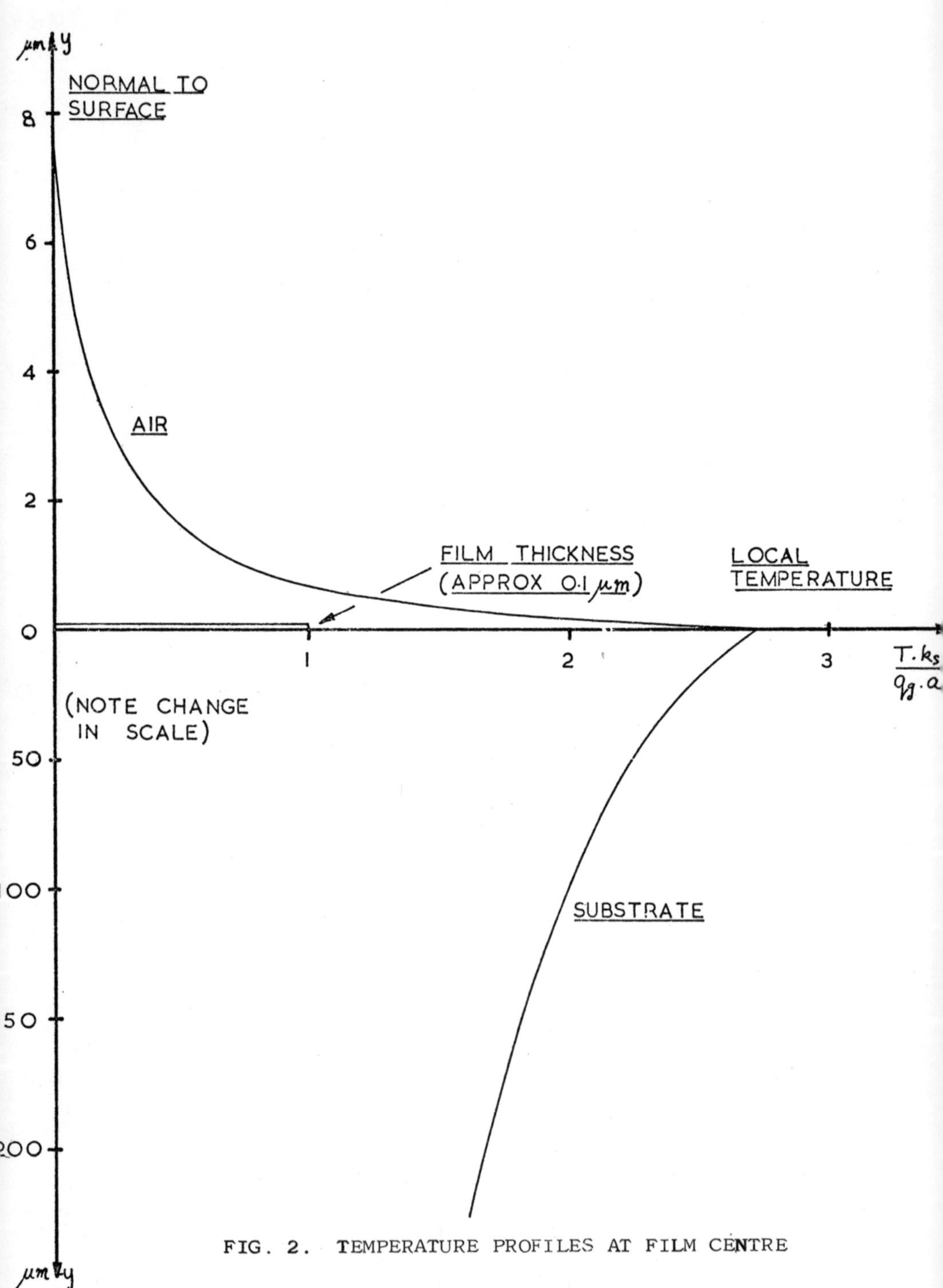

FIG. 2. TEMPERATURE PROFILES AT FILM CENTRE

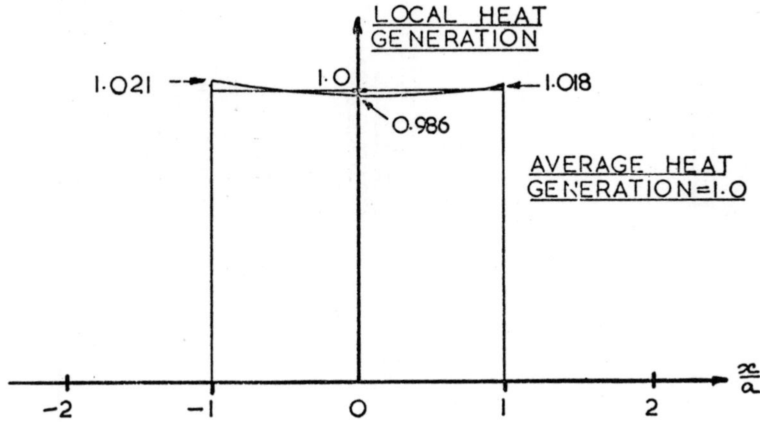

FIG.3. LOCAL HEAT GENERATION

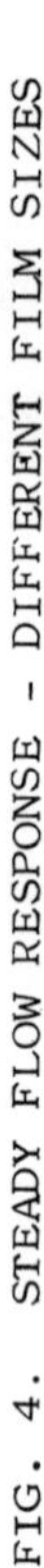

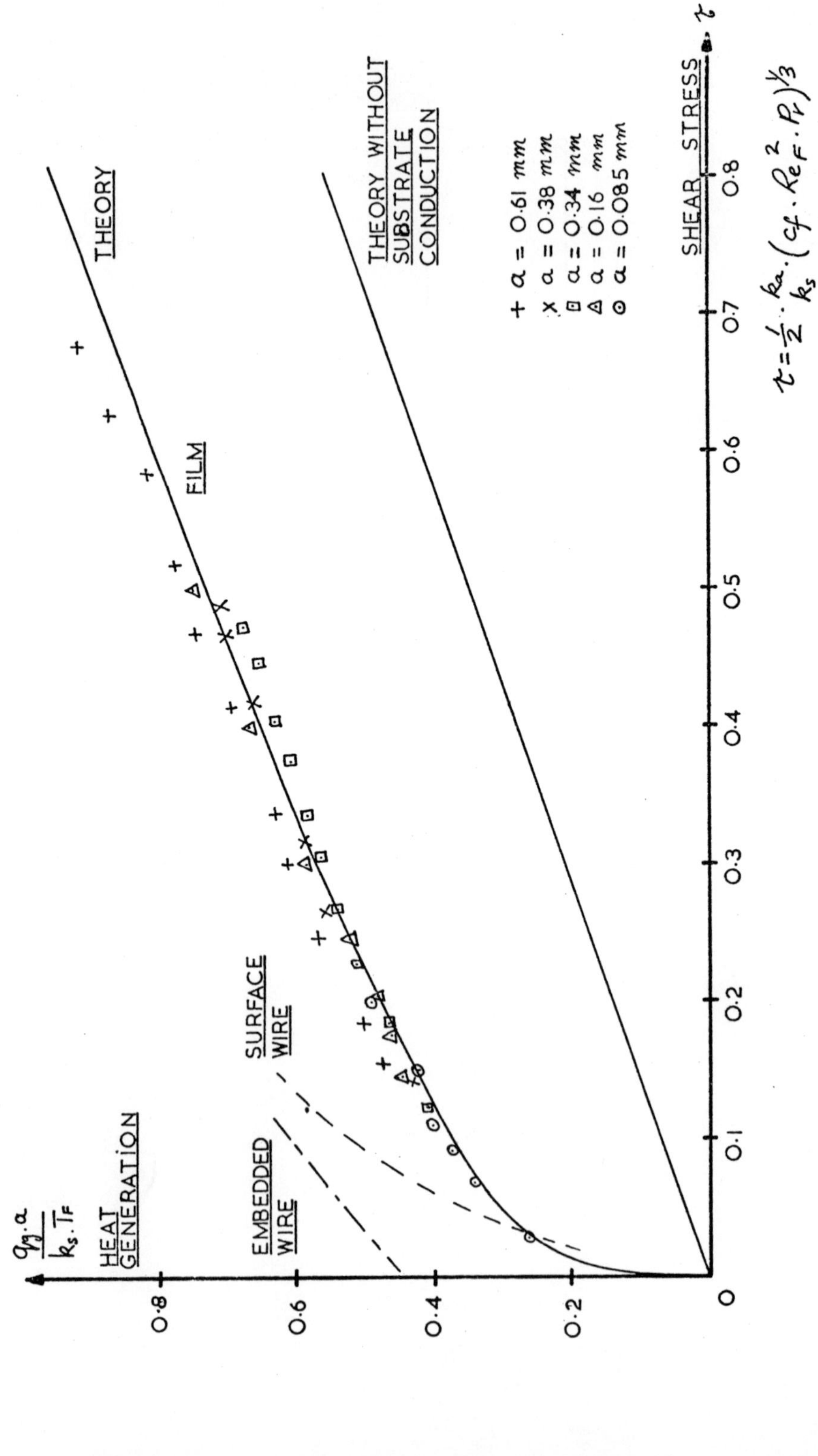

FIG. 4. STEADY FLOW RESPONSE - DIFFERENT FILM SIZES

FIG. 5. STEADY FLOW RESPONSE - DIFFERENT FILM TEMPERATURES

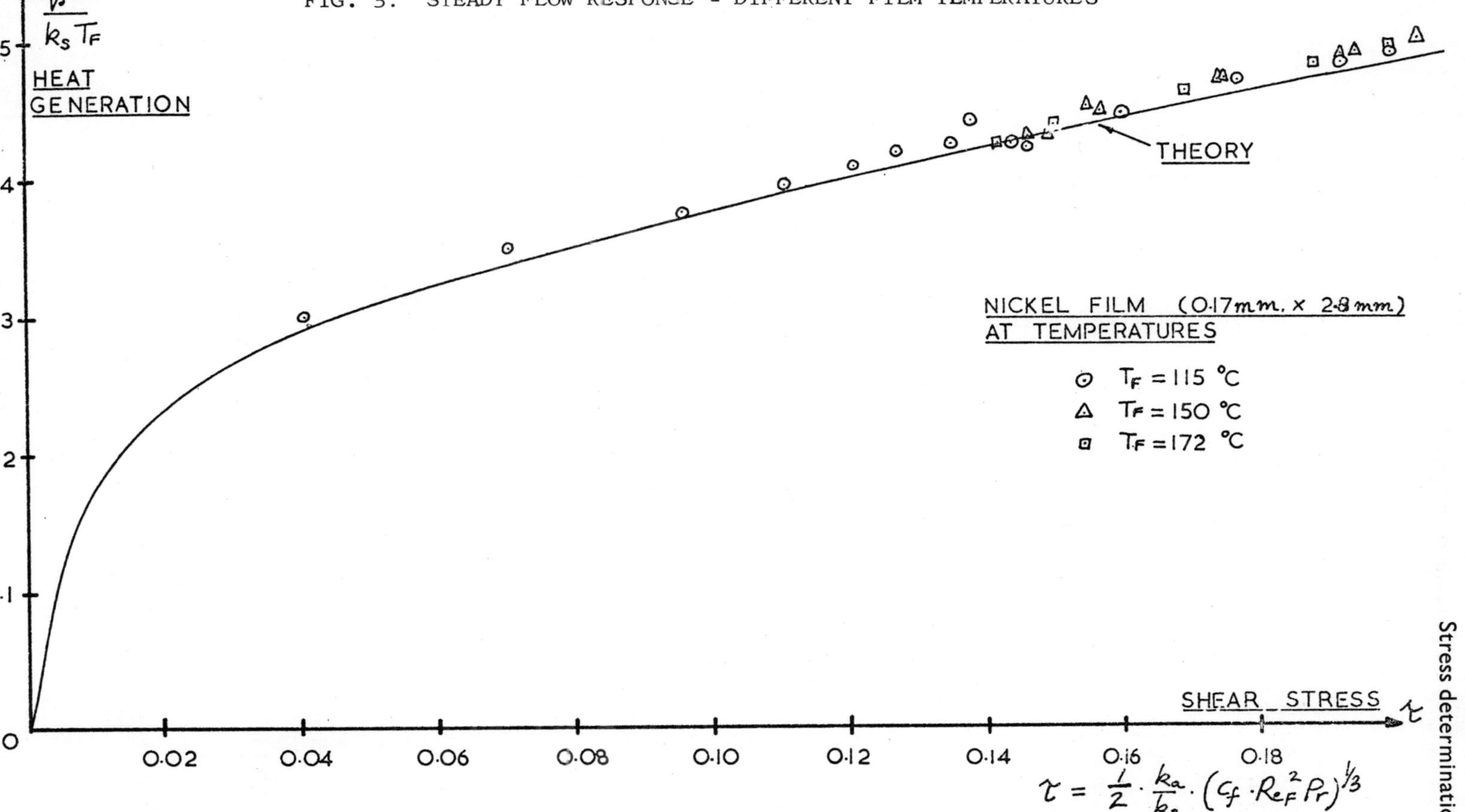

FIG. 6. RESPONSE TO HARMONIC SIGNAL FOR WIRES AND FILM

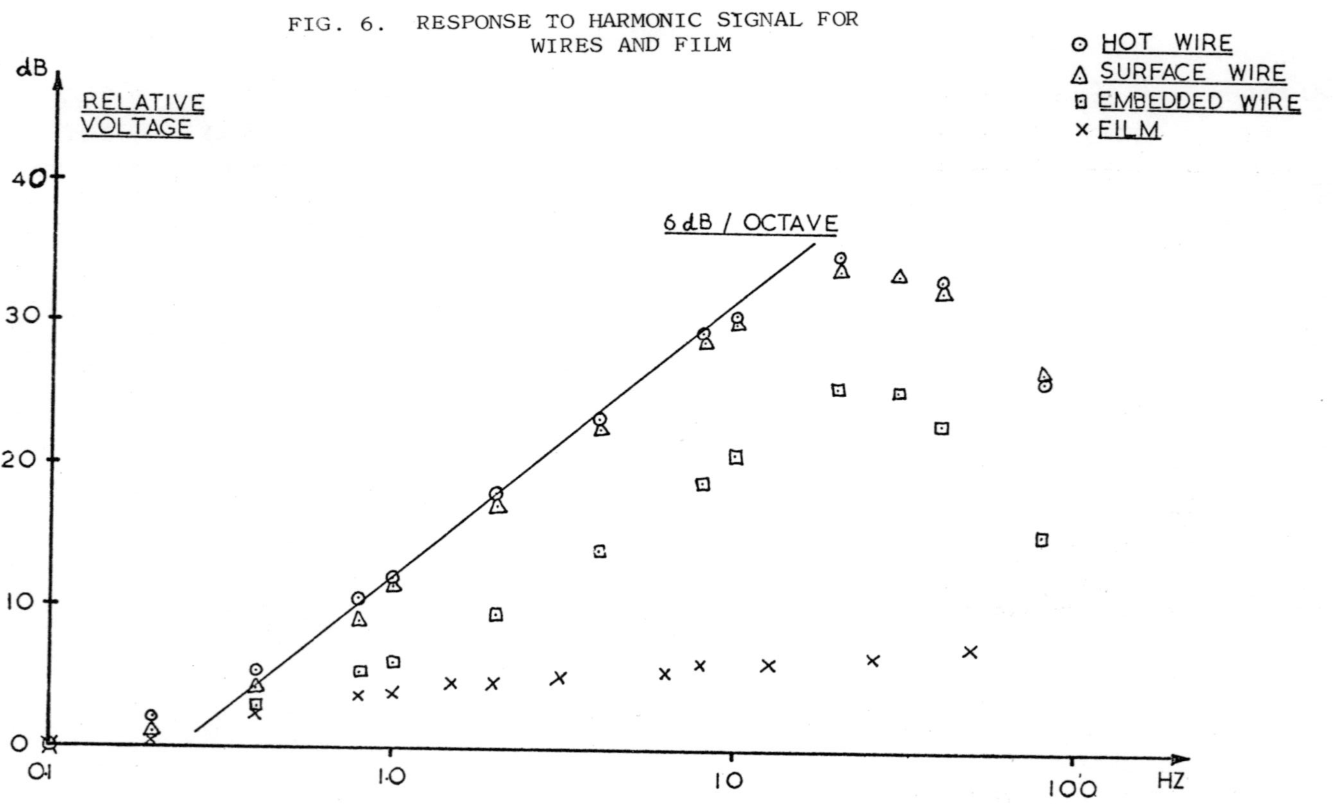

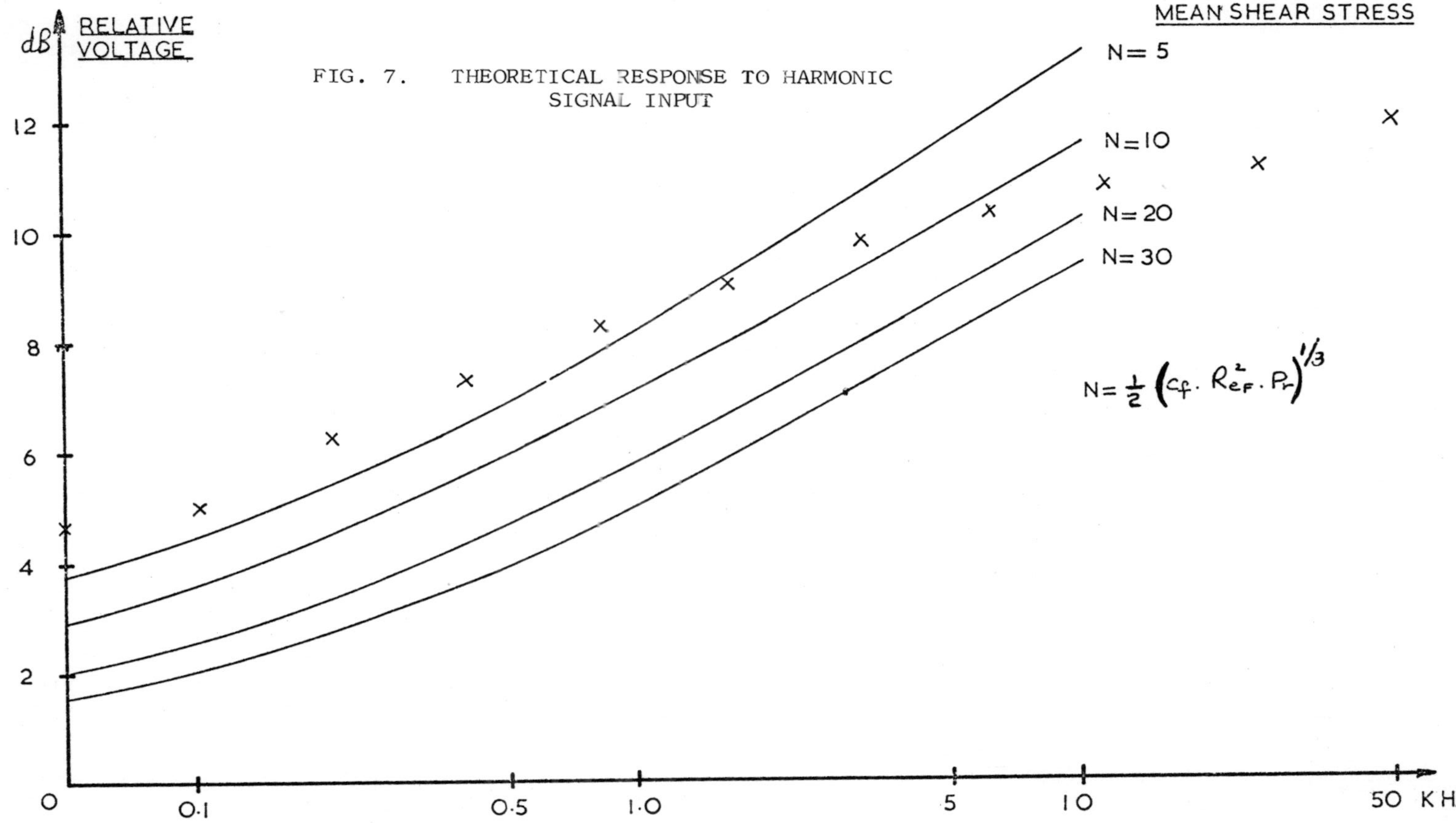

FIG. 7. THEORETICAL RESPONSE TO HARMONIC SIGNAL INPUT

V Measurement of temperature

V-2 Questions addressed to H.P.McNAIR, The Gas Council, London

E.A.SPENCER asks: The author claims high accuracy and the advantage of not requiring intermediate calibration for his instrument. Could he give repeat calibrations to show how long the constants in the equation hold good to within the ± 2 per cent level?

A.AHMAD asks: The author has used two thermistors, one for temperature and the other for velocity measurement. They are displaced in space and therefore the results obtained refer to different conditions prevailing at two points. Is it possible to use a single thermistor which operates in self-heating and temperature sensing modes in the same position?

L.MELION asks: In your final statement you claim an accuracy of 5% at a calibration velocity of only 0.05m/s. Could you give some more details about your calibration device?

H.P.McNAIR replies: To E.A.SPENCER, the graph shows, in the nomenclature used in the Appendix to the paper, three calibrations performed over a time interval of several months.

To A.AHMAD, you are, of course, correct in saying that the temperature and speed sensors measure conditions at two spatially different points. This is, however, only important where there are likely to be significant spatial gradients, and this is not the case in our study of room environments where the objective is to establish a fairly uniform microclimate. Where gradients are experienced it should be possible to position the sensors closer together in such a way that the effects of the gradients are minimised, or, as Mr Ahmad states to use a single sensor to measure both temperature and speed. This latter would, however, involve the use of switching circuity to apply one voltage to the sensor when it is used as an unheated sensor to measure temperature, and a higher voltage when it is used in the self-heating mode to measure speed. This would negate one of the advantages of the instrument in that it would not allow such a simple measurement of both parameters by sequential and automatic scanning via a data-logger.

To L.MELION, the accompanying table of computer calibration data may be useful in answering Mr. Melion's question. The right-most column gives the percentage difference between the measured values of speed and those calculated from the computed equation. The calibrations were performed on a whirling-arm rig which is similar to that described in reference (2) to the paper.

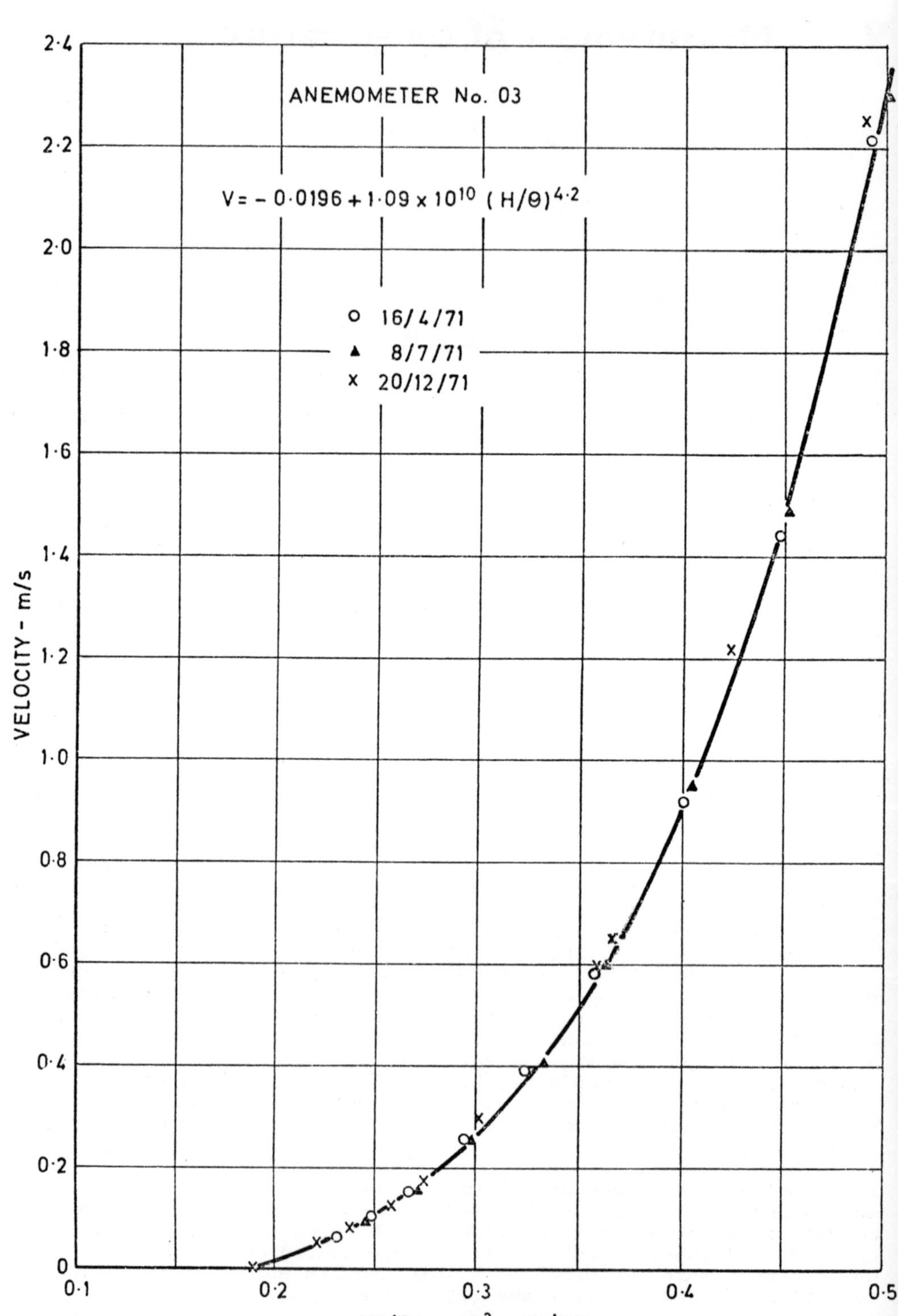
CALIBRATION CURVE OF AN ANEMOMETER
ANEMOMETER No. 03
V= – 0·0196 + 1·09 x 10^10 (H/θ)^4·2
○ 16/4/71
▲ 8/7/71
x 20/12/71
VELOCITY - m/s
2·4
2·2
2·0
1·8
1·6
1·4
1·2
1·0
0·8
0·6
0·4
0·2
0
0·1
0·2
0·3
0·4
0·5
(H/θ) x 10^2 – W/ °C

```
V = -0.348E-01 + 0.923E 11 * (H/THETA) ** 4.5
CORRELATION COEFF. = 0.9999
```

H/THETA	V.ACT m/s	V.CALC m/s	DIFF m/s	PCT.DIFF %
0.183E-02	0.000	0.010	0.010	100.000
0.221E-02	0.066	0.069	0.003	4.245
0.235E-02	0.101	0.102	0.001	0.762
0.277E-02	0.248	0.251	0.003	1.360
0.302E-02	0.385	0.387	0.002	0.612
0.328E-02	0.585	0.577	-0.008	-1.307
0.360E-02	0.910	0.896	-0.014	-1.557
0.397E-02	1.421	1.411	-0.010	-0.737
0.438E-02	2.201	2.214	0.013	0.611

VII Medical measurements by thermistor and hot-film probe techniques

VII-1 Questions addressed to W.M.MORRIS and J.MURRAY, University of Edinburgh and J.H.FILSHIE, Poultry Research Centre Edinburgh

A.DITTMAR asks: Is it indiscreet to ask the exact composition of the metal oxides mixture used for the thermistor?

How much is the maximum temperature difference between thermistor and blood during a pulse of heat?

W.M.MORRIS, J.MURRAY and J.H.FILSHIE reply: The thermistors are produced by d.c. sputtering of a sintered target consisting of

50% Manganese Oxide
25% Nickel Oxide
25% Copper Oxide

in an Argon/Oxygen atmosphere in the ratio 3:1. The composition of the thermistor is not identical to that of the target. X.R.F. analysis indicates that there is a lower proportion of Mn in the film than in the target and a slight increase in Nickel over Copper.

The maximum instantaneous temperature difference is 10°C lasting for 50mS, at intervals of approximately 1 sec.

It is, however, misleading to consider only this temperature difference under transient conditions and a more realistic indication is the average heat transfer into the medium.

Any calculations must be approximate, but we estimate that the system described injects an average power of $\frac{1}{2}$ mW into the fluid. This would imply, for a flow rate of 1mℓ/sec, a temperature rise of 0.0001°C over 1 sec.

The absence of coagulation in our experiments indicates that the effective surface temperature has negligible effect on the blood.

VII-2 Questions addressed to W.A.SEED, Imperial College London and I.R.THOMAS, Kings College Hospital, London

D.E.M.TAYLOR asks: What is the frequency characteristic for pressure determination using your probe? Unless this is also good the potential use of your probe in the estimation of rate of energy production etc. would be severely limited.

W.A.SEED and I.R.THOMAS reply: The pressure frequency response, as with all catheter - manometer systems depends largely upon the catheter dimensions and wall-properties. In this work the catheters were connected to Statham P23 Gb pressure transducers; with care taken to eliminate bubbles these assemblies usually had a resonant frequency between 60 and 80 Hz and were underdamped (damping ratio 0.2) therefore they would record faithfully (i.e. amplitude within 5%) to somewhere in the region of 20 Hz and phase lags would be small.

In this work we have not so far used pressure measurements to derive other parameters and for most clinical purposes the performance described would be acceptable. If greater precision were required the laborious procedure of Fourier analysis of the pressure wave, followed by phase and amplitude correction of each harmonic team and re-synthesis of the corrected waveform would be necessary.

Written contribution received from D.S.TUNSTALL PEDOE

We have been making thin film blood velocity measurements in Oxford in animals and at the time of cardiac catheterisation and heart valve replacement in man since 1967.

Dr. Francis commented in his paper that his group had not found evidence of reverse flow near the wall of the descending aorta when traversing this vessel with their thin film probe, but they were not able to make measurements close to the wall. On theoretical grounds this reverse flow near the wall would seem unlikely in the presence of a fairly flat profile for forward flow. We can confirm the work of Ling[1] in this respect[2] and I think the reason that reverse flow is seen near the wall of the aorta even though the forward blood velocity across the vessel is uniform is because of the elasticity of the aorta which unloads in early diastole. Since it is stiffer more peripherally there is a reverse flow wave that can be seen on electromagnetic flow records and possibly because of the proximity of the unloading vessel walls appears to occur only at the edges of the vessel.

Dr. Francis's velocity profiles are very similar in other respects to the ones that we recorded.

I found the paper by Seed and Thomas very interesting. I was intrigued to see in the Abstracts that they described their technique of velocity measurements in the ascending aorta of man with a thin film catheter as a relatively non-invasive technique. As a cardiologist I must quibble with this use of the term. Any catheterisation of the ascending aorta carries a definite mortality and some risk to the artery used. I would also question whether any patient undergoing cardiac catheterisation has a 'normal' heart, even if no abnormalities were found during the cardiac catheterisation. Only gross abnormalities of blood flow are shown by catheterisation procedures and angiography and the presence of turbulence in the ascending aorta in such a patient should I think not be taken as meaning that it is a normal occurrence under resting conditions.

Finally if I might just comment on the value of blood acceleration measurements for ascending aortic blood flow. I think it would be unfortunate if this became a general measurement in cardiology without further thought as to its physical meaning. The acceleration measured will depend on the flow rate and the cross-sectional area of the ascending aorta. As Dr. Schultz mentioned in his paper, the geometry of the situation is often the most difficult factor to measure in medical fluid dynamics and since there may be wide variations in the diameter of the aorta of patients of different ages for the same flow rates, comparison of acceleration without the other measurements may be meaningless. Perhaps blood acceleration time will be more useful.

References.

1. S.C.Ling, H.B.Atabek, D.L.Fry, D.J.Patel and J.S.Janicki, "Application of heated-film velocity and shear probes to haemodynamic studies" Circulation Research 23: 789-801.

2. D.S.Tunstall Pedoe, "Velocity distribution of blood flow in major arteries of animals and man" D.Phil.Thesis. Oxford University (1970).

W.A.SEED and I.R.THOMAS reply: Catheter-tip hot-film probes offer one exclusive advantage; they permit direct examination of aortic flows for the presence of turbulence, and should settle speculation about its occurrence in the circulation which has gone on for many years. In addition, it offers the possibility of assessing the time-course of venticular ejection; this may be of value in the clinical assessment of the heart function. Thus study of a wide range of disease states in a clinical investigation laboratory will both evaluate the instrument as a clinical tool, and provide information of physiological and fluid dynamic interest.

Discussion on the utility of acceleration measurements.

A claim made during Dr. Thomas's presentation that aortic acceleration measurements were valuable in assessing the condition of patients' left ventricle was challenged by Dr. Tunstall-Pedoe. He argued that dimensional information and the duration of acceleration must be taken into account. Dr. Seed replied that in the absence of evidence to the contrary, the usefulness of acceleration as an index should not be dismissed.

L.H.LIGHT then contributed: That changes in acceleration in any one individual reflect changes in ventricular contractile force has been repeatedly verified. Unfortunately, however, it is not possible to assess the condition of a patient's ventricular myocardium from acceleration measurements alone, as they are influenced (in addition to the factors mentioned by Dr.Tunstall-Pedoe) by the compliance of his vasculature and by the hormonal and nervous drive to his heart*. We have seen patients in extreme left heart failure whose aortic acceleration was twice as high as that of an athlete at rest.

*G. Cross and L.H.Light : "Cardiovascular Information from Transcutaneous Aortovelography" Proc.3rd Internat.Conf.on Med. Physics, Gothenburg, Aug. 1972, in press.

VII-3 Questions addressed to G.P.FRANCIS, K.M.KISER and H.L.FALSETTI, State University of New York at Buffalo, U.S.A.

D.E.M. TAYLOR asks: Using a Pitot-tube traverse we have observed flow reversal close to the wall (about 2mm) before reversal of axial flow, but this has been confined to dogs with a low cardiac output. Have you observed similarly?

D.R. BROUGHNER asks: From aortic model studies we have seen cross channel steaming downstream from branch arteries. This could explain varying velocity profiles. Have you made measurements in various planes at the same point in the aorta?

G.P.FRANCIS, K.M.KISER and H.L.FALSETTI reply: To D.E.M. TAYLOR, we are currently looking at the details near the wall using a new 0.5 conical probe. As we note in our manuscript and on the figures, ". . .the 1.5 mm probe was too large to make precise measurements in the wall regions. . . ". We will be particularly attentive to your comment for dogs with low caridac output.

To D.R. BROUGHNER, yes, but only in the ascending aorta. Currently we are making these measurements in both the ascending and descending aorta using a new 0.5 mm diameter conical probe. As we note on page 308 of Volume I: "In the descending aorta the variations may be the results of branching. . . ".

VIII Medical measurements by pressure transducer and doppler techniques

VIII-1 Questions addressed to D.S.TUNSTALL PEDOE, Radcliffe Infirmary, Oxford

P.LADBROOK asks: Have fluid dynamic properties been obtained for the prosthetic tricuspid valve developed by the Biomechanical Engineering team at Sussex University, either in vivo, as for the Starr-Edwards device, or in vitro?

D.S.TUNSTALL PEDOE replies: I have not been involved with the work from Sussex University and so I am afraid that I cannot answer your question. I have assessed the competence of homograft and heterograft biological valves as well as Starr-Edwards prosthetic ball valves using this technique and one of its chief uses might be the in-vivo assessment of these valves which sometimes become progressively incompetent.

So far as I know no one else has been using this technique for the in-vivo assessment of prosthetic valves, although I believe it was attempted using the brachial artery but without much success. If the Sussex team have tested their valves in-vivo they would probably have used more conventional techniques.

Further written comment by D.S.TUNSTALL PEDOE.

Since preparing the published part of this paper I have been able to look at two of the factors that I have mentioned in more detail and so assess more precisely therein contribution to the difficulties of obtaining quantitative data from this technique.

First from measurements made on angiograms (X rays with the blood vessels made visible by the injection of contrast media) I have been able to derive an approximate value for the cubic capacity of the arterial tree between the aortic valve and the site at which I make the Doppler blood velocity measurements in the subclavian arteries. In the normal patient this works out at about 75 ml. but in patients with aortic valve stenosis with large dilatation of the ascending aorta the figure may be as high as 400ml. I have not been able to obtain figures for the distensibility of the arterial tree under these conditions but it seems likely that there would be a large difference in the capacitance of the proximal arterial tree. Thus the fraction of reverse flow (regurgitant fraction of total forward flow) seen in the subclavian artery as a function of the same measurement in the ascending aorta may vary considerably from patient to patient.

Secondly I have been tape recording the Doppler shift audio frequency signal from the apparatus and passing it through a sound spectrograph. (Kay Elemetrics Co. New Jersey, U.S.A.). This displays a three-dimensional plot of frequency against time, with the intensity of the frequency shown by the depth of the shading. Two examples are shown. The first is taken from a patient with an irregular heart beat and only a slight degree of aortic regurgitation. The Doppler velocimeter produced a good record in which the systolic velocities appeared to be faithfully represented. Three heart beats are shown in the figure (fig. 1). During systole there is an easily seen forward flow wave of blood flow in which the blood velocity appears to have been fairly uniform within the beam of the ultrasound probe. This can be inferred from the narrow band of frequencies shown which outline the waveform. It seems likely from this record that the blood flow in the subclavian artery had a fairly uniform velocity since a parabolic velocity distribution within the vessel would produce an even intensity of frequencies with a wide band width. It is noticeable too that the systolic upstroke of velocity is well shown, unlike fig. 2.

Figure 2 is the spectrograph of the Doppler signal from the subclavian artery of a patient with severe aortic incompetence in whom a record similar to that shown in fig. 6 on page 331 of Vol.I was obtained. Systole was poorly represented and there was obvious preponderance of diastolic reverse flow velocity as a result. In this spectrograph systole overlies the caption and diastole extends from the caption diastole to the S of systole. The onset of systole is marked by a large low frequency thump which probably corresponds to an artefact from the Doppler shift produced by the rapid dilatation of the artery. This signal is so intense that the simultataneous Doppler signal from the moving red blood cells seems to have been lost. This fall-out of the signal at the onset of systole is followed by a systolic trace showing a much greater bandwidth of frequencies than that of fig. 1. The consequence of the spreading of the intensities of the Doppler shift frequencies, so that a bandwidth of noise is presented to the zero crossing detectors of the Doppler velocimeter, would be the failure of this device to give a representative output. In diastole however the frequencies shown by the spectrograph still show some high frequency emphasis and it is therefore not surprising that the gross reverse flow in diastole is well demonstrated.

The essential diagnostic value of this technique is therefore maintained and even if it never becomes more quantitative, by improvements in the apparatus and our ability to compensate for the variables in different patients, it should prove a useful aid to the cardiologist.

VIII-2 Questions addressed to L.H.LIGHT, Clinical Research Centre, Harrow

M.T.THEW asks: What arrangements are made to distinguish electronically between positive and negative Doppler shifts when examining blood velocity in the region of the aortic arch?

L.H.LIGHT replies: A heterodyne system is used as outline in my paper under the heading "Direction-resolving Demodulation". This solution* to the problem of discriminating against flow in the branch arteries was preferred to the alternative possibility of range-gating, as (i) it dispenses with the need for an additional control to select the depth at which the flow is most closely in line with the ultrasonic beam, and (ii) it sidesteps the extra complexities required to avoid the range and spectral ambiguities to which the basic range-gated system is subject when used at the required depth.

Figure 1. Sound spectrograph of Doppler sound frequencies from subclavian artery of patient with mild aortic incompetence. Note the high frequency outlining of the systolic waves.

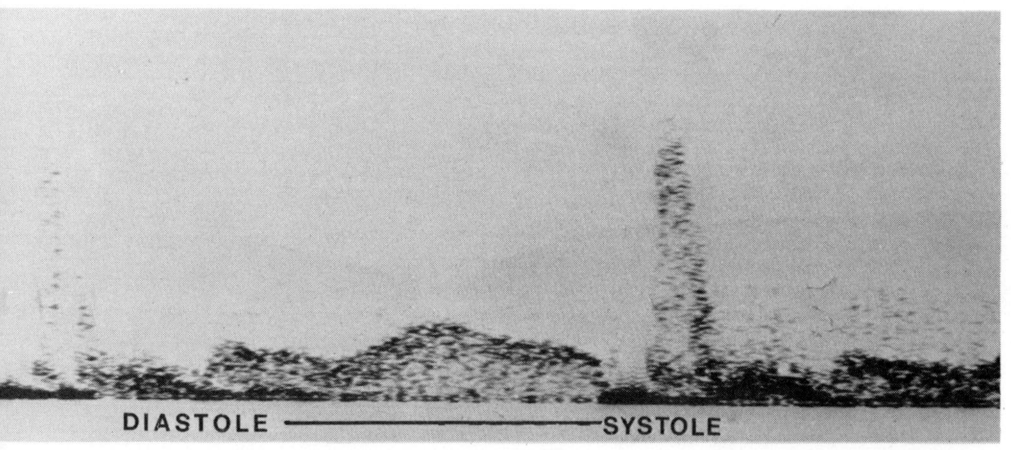

Figure 2. Sound spectrograph of Doppler sound frequencies from subclavian artery of patient with severe aortic incompetence. Systole is initiated by a low frequency "thump" and a mixed band of frequencies follow. Diastole is well marked by high frequency outlining of the waveform.

* L.H.Light and G. Cross : "Cardiovascular Data by Transcutaneous Aortovelography" in "Blood Flow Measurement", Ed.V.C.Roberts, Sector Publishing Ltd., London, 1972.

A Question addressed to D.S.TUNSTALL PEDOE, L.H.LIGHT, D.E.M. TAYLOR and J.S. WHAMOND by M.J.BENNETT:

Using ultrasonic techniques, how do you ensure that no blood, tissue or chromosome damage will be caused by the power ratings used? Are there any definitive papers in this area?

D.S.TUNSTALL PEDOE replies: There is much work being done in this field using a variety of biological preparations, ultrasound frequencies and intensities. Theoretically, damage could arise from cavitation, plasma streaming and disruption of intra-cellular particles. There has been one report of white cell chromosome damage to blood left in contact with a continuous wave Doppler probe for some hours. Despite repeated attempts by other workers no one has been able to repeat the results of this experiment and it now seems likely that the frequencies and power levels used in Doppler devices are without observable effects on mature tissue. Higher energy levels cannot be tolerated because of the local heating of the tissue. This is just perceptible with some commercial devices.

L.H.LIGHT replies: While damage to tissues can readily be demonstrated at power levels some 100 times higher than those used in Doppler examinations, it is, in the nature of things, vastly more difficult to be sure of the absolute safety at practical levels ($\sim$20mW/cm^2). No immediate harmful effects have been seen in man or animal studies. The same applies (with one doubtful exception[1] which has, however, not been confirmed by similar experiments elsewhere [2,3]) to studies in human tissues. Follow-up studies extending to 7 years on children examined in utero have also failed to reveal any abnormality attributable to exposure to ultrasound. Intermediate power levels ($\sim$100mW/cm^2) have been shown to accelerate wound healing[4].

While this does not add up to absolute certainty, it is reasonable to conclude that any risk involved is probably less than that associated with alternative methods, or that of doing without the information obtained by Doppler investigation. A useful review article has also been published[5]. For up-to-date information two forthcoming publications should be consulted[6,7].

(1) I.MacIntosh and D.Davey: "Chromosome aberrations induced by Ultrasonic Fetal pulse detector", Brit.Med.J., 1970(4), pp.92-93.

(2) W.T.Coakley et al: "Chromosome Aberrations after Exposure to Ultrasound", Brit.Med.J., 1971(1), pp.109-110.

(3) U.Abdulla et al: "Effect of diagnostic Ultrasound on Chromosomes", Lancet, 2, No:7729, Oct.71, pp.829-831.

(4) M.Dyson et al: "Stimulation of Tissue Regeneration by Ultrasound", Clin.Sci., 35 (1968), pp.273-285.

(5) C.R.Hill: "Possibility of Hazard in Med.and Ind.Applications of Ultrasound" Brit.J.Radiol., 41 Aug. 1968, pp.561-569.

(6) C.R.Hill in "Clinical Aspects of Ultrasonic Diagnosis" Ed. P.N.T.Wells, Churchill, London, 1972, Chapter 8 (in press).

(7) Brit.J.Radiol., May 1972 issue (collection of relevant papers).

The Disa Grant

RULES

DISA, the Dansk Industri Syndikat A/S, a company which make instruments for fluid dynamic measurements, have agreed to make a single grant to the value of up to £1000 to further research and development within the field of experimental fluid dynamics.

Applications are therefore invited from those who would wish to be considered for this Grant. No restriction is placed on the way in which it is proposed to spend the grant, for example using it in association with other funds, allocating it to some independent project or spending it on travel to some centre are all possible proposals. Similarly there is no restriction on type of project described which might well be of a research or of a development character. The successful applicant will have demonstrated to the international selection committee that the scheme proposed is practicable, original and timely.

Applications must be submitted as follows:

1. They must take the form of a typed letter in English, addressed to the Secretary of the Disa Grant Committee, Disa, DK 2730, Herlev, Denmark, to reach him by 1st July 1973. Letters exceeding 1000 words in length will be rejected. No other correspondence should be entered into at this stage.

2. The letter should clearly describe the specific fluid dynamic purpose to which the grant would be put and explain its practicability originality and timeliness. It should also state the intention and place of execution of the project. If necessary, reference should be made to relevant publications.

3. The names and addresses of two persons to whom reference may be made as to the suitability of the applicant for the work proposed and (where relevant) the possibility of completing the project in the place described should be included with the letter.

4. Short-listed applicants may be required in due course to submit further details of their projects to the committee.

5. The successful candidate will receive the grant in not more than two instalments, the final one being dependent upon submission of a satisfactory progress report to the Grant Committee. This report may be published in due course in Disa Information and agreement so to publish is a prior condition of the Grant.

6. The Grant Committee may only be approached through their Secretary. All decisions it may make are final.

THE GRANT COMMITTEE

Chairman : Dr. D.J. Cockrell,
University of Leicester, Leicester, U.K.

Professor G. Comte-Bellot,
Laboratoire de Mecanique des Fluides, Lyon, France.

Professor R.E. Kaplan,
University of Southern California, Los Angeles, U.S.A.

Professor E. Naudascher,
Institut fur Hydromechanik, Karlsruhe, Germany.

Professor K. Refslund,
Danish Technical Institute, Lyngby, Denmark.

Secretary : Mr. F.V. Steenstrup,
Disa Elektronik A/s, Herlev, Denmark.

The Disa Conference
University of Leicester 12-14 April 1972

COMMITTEE

R. Barber	Land Pyrometers Ltd
C. Boorman	U.K.A.E.A. (Risley)
D. J. Cockrell	University of Leicester
A. L. Cussens	Disa
R. A. E. Fursey	Shell Research (Egham)
D. L. Schultz	University of Oxford
E. A. Spencer	National Engineering Laboratory
J. C. Tipping	Gas Council (Fulham)
H. S. Wolff	Bio-Engineering Division, Clinical Research Centre

NAMES AND ADDRESSES OF DELEGATES

AHMAD, A.
Fisons Ltd., Fertiliser Division,
Levington Research Station,
Technical Development Dept.,
Levington,
Ipswich, Suffolk, 1P10 0LU

AMOS, P. G.
Disa,
116 College Road,
Harrow, Middx, HA1 1HQ

ARQUES, Dr. P.
Laboratories des Mecanique,
272 Avenue des Allies,
51 Chalons sur Marne, France

ARROWSMITH, Dr. A.
Sheffield University,
Dept. of Chem. Eng. & Fuel Technology,
Newcastle Street, Sheffield, S1 3JD

BAKER, R. J.
A.E.R.E., U.K.A.E.A.,
Harwell, Didcot, Berks.

BALAFOUTAS, G. J.
15 Vassilissis Olgas,
Thessaloniki, 362, Greece

BARBER, R.
Land Pyrometers Ltd.,
Wreakes Lane, Dronfield,
Sheffield, S18 6PN

BEAN, R.
Electrical Research Association,
Cleeve Road, Leatherhead, Surrey

BENNETT, M. J.
Brunel University,
Dept. of Mech. Eng.,
Kingston Lane,
Uxbridge, Middx.

BIALOSTOCKI, Dr. S. J.,
Polish Academy of Sciences,
Institute of Fluid Flow Machines,
ul. Fiszera 14 Gdansk, Poland

BIRCH, A. D.
Gas Council,
Midlands Research Station,
Solihull, Warwicks.

BLAKE, K. A.
National Engineering Laboratory,
East Kilbride, Glasgow

BLOM, Mr.
Koninkliijke/Shell,
Laboratorium Amsterdam,
Badhuisweg 3,
Amsterdam-Noord, Holland

BOORMAN, C.
U.K.A.E.A.
Risley Eng. & Materials Lab.,
Risley, Nr. Warrington, Lancs.

BOUGHNER, Dr. D. R.
Hammersmith Hospital,
Royal Postgraduate Medical School,
Dept. of Clinical Cardiology,
London, W12

BOURKE, P. J.
U.K.A.E.A.,
Room 1.85, Building 329,
A.E.R.E., Harwell,
Didcot, Berks.

BOUTIER, A.
Office National D'Etudes et de Recherches Aerospatiales,
29 Avenue de la Division
Leclerc, 92 Chatillon, France

BOYE, J. Secher,
Disa, Herlev DK 2730,
Denmark

BRADLEY, Dr. C. I.
Gas Council,
Midlands Research Station,
Solihull, Warwicks.

BRADSHAW, P.
Imperial College,
Dept. of Aeronautics,
Prince Consort Rd., London, SW7

BRAND, Dr. R. S.
I.S.V.R., The University,
Southampton, SO9 5NH

BROSH, A.
Technion,
Israel Institute of Technology,
Haifa, Israel

BROWN, Dr. D. R.
Gas Council,
Midlands Research Station,
Solihull, Warwicks.

BRUUN, Dr. H. H.
I.S.V.R., The University,
Southampton, SO9 5NH

BULLOCK, Dr. L. E.
The Foxboro Company,
38 Neponset Avenue, Foxboro,
Mass. U.S.A. 02035

BURTON, R. C.
UWIST,
Dept. of Mech. Eng. & Eng. Prod.,
King Edward VII Avenue,
Cardiff, CF1 3NU

BUUR, E. J.
TNO-IWECO,
Delft, Holland

CHEESEWRIGHT, Dr. R.
Queen Mary College,
Dept. of Mech. Eng.,
Mile End Road, London, E1

CHIGIER, Dr. N. A.
University of Sheffield,
Dept. of Chem. Eng.,
Newcastle St., Sheffield, SL 3JD

CHOMETON, F.
CNAM, 21 rue Pinel,
Paris 13^{e}, France

CHRISTENSEN, H.
Svenska DISA AB, Visattravagen 41,
141 49 Huddinge Sverige, Sweden

CHRISTESEN, G.
DISA Electronique SARL, P.B. 22,
94 Villeneuve-le-Roi,
France

COCKRELL, Dr. D. J.
University of Leicester,
Engineering Department,
Leicester, LE1 7RH

CUSSENS, A. L.
Disa, 116 College Road,
Harrow, Middx. HA1 1HQ

DAMION, J. P.
Conservatoire National des Arts
et Metiers,
21 rue Pinel, Paris 13^e, France

DAVENPORT, C. J.
Associated Engineering Development,
Cawston House, Cawston, Rugby

DAVIES, Dr. P. A. O. L.
I. S. V. R., The University,
Southampton, SO9 5NH

DELHAYE, Dr. J. M.
C. E. A.,
Centre de Grenoble (38), France

DITTMAR, A.
Faculte de Medicine et de Pharmacie
de l'Universite de Lyon,
8 Avenue Rockefeller, Lyon (8^e),
France

DUFAUX, J.
GPSP Universite Paris VII,
2 Place Jussieu,
Paris 5^e, France

DUFOUR, A.
c/o Asgen, Via N. Lorenzi 8,
16152 Genova, Cornigliano, Italy

DURAO, D.
Imperial College,
Dept. of Mech. Eng.,
Exhibition Road, London, SW1

DURST, F.
Imperial College,
Dept. of Mech. Eng.,
Exhibition Road, London, SW1

DVORAK, K.
Sheffield University,
Dept. of Chem. Eng. & Fuel Tech.,
Newcastle Street, Sheffield S1 3JD

ELBERG, S. M.
Commissariat a l'Energie Atomique,
B. P. 510,
29-33 Rue de la Federation,
Paris 15^e, France

ELSNER, Dr. J. W.
Institute of Heat Machinery,
Technical University of Czestochowa,
Al Zawadzkiego 21,
Czestochowa, Poland

ELSWORTH, J. E.
Shell Research Ltd.,
Egham Research Laboratories,
P. O. Box 11, Egham, Surrey

FAIRBANKS, F.
Hatfield Polytechnic,
Hatfield, Herts.

FAZEKAS, I.
AUTOKUT, Budapest,
XI Bartok B. ut 104,
Hungary

FEATHERSTONE, W. B.
British Steel Corporation,
Research Centre, Commerce Way,
Skippers Lane, South Bank,
Teesside, TS6 6UT

FEW, Dr. P. C.
Leicester Polytechnic,
The Newark, Leicester

FILSHIE, J. H.
University of Edinburgh,
Wolfson Microelectronics Liaison
Unit, School of Eng. Science,
Mayfield Road, Edinburgh, EH9 3JL

FLOWER, Dr. K. A.
Salford University,
Dept. of Mech. Eng.,
Salford, Lancs. M5 4WT

FRANCIS, Prof. G. P.
State Un. of New York at Buffalo,
Dept. of Mech. Eng.,
Faculty of Eng. & Applied Sciences,
New York 14214, U. S. A.

FREYTAG, C.
Meteorologisches Ins. der Universitat Munchen,
8 Munchen 80, Montgelasstr. 3,
W. Germany

FRIEND, P. G.
Ferraris Development & Engineering Co. Ltd.,
26 Lea Valley Trading Estate,
Angel Road, London, N18 3JD

FURSEY, R. A. E.
Shell Research Ltd.,
Egham Research Laboratories,
Egham, Surrey

GALE, Dr. B. C.
Disa, 116 College Road,
Harrow, Middx. HA1 1HQ

GAUDET, M.
GPSP, Universite Paris VII
2 Place Jussieu,
Paris 5^{e}, France

GEIGER, D.
GPSP, Universite Paris VII
2 Place Jussieu,
Paris 5^{e}, France

GEORGE, P. T.
British Nuclear Design and Construction Ltd.,
Cambridge Road, Whetstone,
Leicester LE8 3LH

GEORGE, Dr. W. K.
Pennsylvania State University,
1400 Circleville Road,
PA 16801 U.S.A.

GIBSON, Dr.
University of California,
San Diego, U.S.A.

GIRGIS, Dr. N. S.
Lanchester Polytechnic,
Dept of Mech. Eng.,
Eastlands, Rugby, Warwicks.

HALLIWELL, G.
U.K.A.E.A., The Reactor Group,
Reactor Fuel Element Labs.,
Springfields, Salwick,
Preston, Lancs.

HARDING, N.
Lanchester Polytechnic,
Dept. of Mech. Eng.,
Faculty of Engineering,
Coventry, Warwicks.

HARVEY, J.
Rolls Royce & Associates Ltd.
P. O. Box 31, Derby

HERBECK, Dr. Margot
Max-Planck-Institut fur Stronmungs-Schung,
34 Gottingen Bundes Republik Deutschland,
Bottingerstr. 6, Germany

HOBSON, B. D.
Lanchester Polytechnic,
Dept. of Mech. Eng.,
Priory Street, Coventry, CV1 5FB

HODSON, R. E.
U.K.A.E.A.
The Reactor Group,
Reactor Fuel Element Labs,
Springfields, Salwick,
Preston, Lancs.

HOELGAARD, J. C.
Disa Electronik A/S,
Herlev DK 2730, Denmark

HOFFMEISTER, Prof. Dr. M.
DAW zu Berlin, Z1 fur Mathematik und Mechanik,
DDR-108 Berlin, Mohrenstrabe 39,
Germany

HUNTER, J. J.
National Engineering Laboratory,
East Kilbride, Glasgow

JENKIN, R.
10 Cranbrook Drive, Kennington,
Oxford, OX1 5RR

JERNQVIST, L.
Chalmers University of Technology,
Dept. of Applied Thermo & Fluid Dynamics,
Fack S-402, 20 Gothenburg 5,
Sweden

JONES, D. H. S.
Disa,
116 College Road,
Harrow, Middx, HA1 1HQ

JOOSTEN, B. P. A.
Twente Institute of Technology,
P. O. Box 217,
Enschede, Holland

KACKER, Dr. S. C.
University of Newcastle,
Dept. of Mech. Eng.,
Stephenson Building, Claremont Rd.,
Newcastle-upon-Tyne, NE1 7RU

KAMOI, A.
The Composite Research &
Development Centre of Toyo Seikan
& Toyo Kohan Companies,
No. 22-4 Okazawa-Cho,
Hodogaya-Ku, Yokohama, Japan

KENDOVSH, A. A.
Birmingham University,
Dept. of Physics,
P. O. Box 363,
Birmingham, B15 2TT

KIELBASA, J.
3400 Gottingen,
Bunsenstr. 14,
Bundesrepublik Deutschland,
West Germany

KIMBER, G. R.
Southampton University,
Southampton, SO9 5NH

KURKI-SUONIO, I.
Tampere University of Technology,
Kouluk 9,
Tampere, Finland

LADBROOK, P.
Disa,
116 College Road,
Harrow, Middx, HA1 1HQ

LAWRENCE, P.
Gas Council,
Watson House,
Peterborough Road,
Fulham, London, SW6

LECROART, H.
S. E. E. N. Bat 104,
CEN SACLAY, B. P. No. 2,
9. 1 GIF Yvette, France

LIGHT, L. H.
Medical Research Council,
Clinical Research Centre,
Watford Road, Harrow, Middx.

LINDE, K.
University of Uppsala,
Dept. of Physical Geography,
Bo. 554, S-751 22 Uppsala 1,
Sweden

LOCKWOOD, J.
Rolls Royce & Associates Ltd.,
P. O. Box 31,
Derby

MABEY, D. G.
Royal Aircraft Establishment,
Bedford

MADSEN, N. J.
Disa Electronik A/S,
Herlev DK 2730, Denmark

MAK, F.
Gas Council, Watson House,
Peterborough Road, London, SW6

MAYE, Dr. J. P.
Lab. Dynamique des Fluides,
40 Av. Rectem Pineau,
86 Poitiers, France

MELION, L.
C/o LUWA AG Anemonenstr. 40
CH 8047, Zurich, Switzerland

MELLING, A.
Imperial College,
Dept. of Mech. Eng.,
Exhibition Road, London, SW7 2BX

McKNIGHT, J.
U. K. A. E. A., R. E. M. L.,
Risley, Warrington, Lancs.

McNAIR, H. P.
Gas Council,
Watson House, Peterborough Road,
Fulham, London, SW6

McWILLIAM, D.
U.K.A.E.A., Reactor Group,
Dounreay, Thurso, Caithness

MINTY, A. G.
UWIST,
Dept of Mech. Eng & Eng. Prod.,
King Edward VII Avenue,
Cardiff, CF1 3NU

MORROW, Dr. T. B.
Loughborough University,
Dept. of Transport Technology,
Loughborough, Leics.

MULLER, A.
Federal Institute of Technology Zurich,
Inst. of Hydromechanics & Water
Resources Management,
8006 Zurich, Tannenstrasse 1,
Switzerland

MURRAY, J.
University of Edinburgh,
Wolfson Microelectronics Liaison Unit,
School of Engineering Science,
Mayfield Road, Edinburgh, EH9 3JL

NALLURI, Dr. C.
Newcastle University,
Dept. of Civil Eng.,
Newcastle-upon-Tyne, NE1 7RU

NASMAN, P.
The Royal Inst. of Technology,
Dept of Aeronautics,
S-100 44 Stockholm 70, Sweden

NIELSEN, P. E.
Disa Electronik A/S,
Herlev DK 2730, Denmark

NIEUWVELT, Dr. C.
University of Technology,
Post Bag 513,
Building W & S Eindhoven,
The Netherlands

NOSKIEVIC Dr. J.
University of Ostrava,
Ostrava 1, tr. Cs Legii 9
Czechoslovakia

PEDOE, Dr. D. S. Tunstall
Radcliffe Infirmary, Cardiac Dept.,
Oxford

PEMIC, A.
Vodogradbeni Laboratorij,
Ljubljana, Jugoslavia

PENOT, Dr. F.
Lab. Dynamique des Fluides,
40 Av. Recteur Pineau,
86 Poitiers, France

PICHON, Dr. M.
Sogreah, Cedex 172,
38 Grenoble Gare, France

PORTE, R.
Commissariat a l'Energie Atomique,
CEN SACLAY, B.P. No. 2
9 GIF sur Yvette, France

PUYAL, C.
Electricite de France,
6 Quai Wattier, 78 Chatou,
France

RAKOVEC, Dr. S.
Prazakova 11, Ljubljana,
Yugoslavia

RIZZO, J. E.
Southampton University,
Southampton, SO9 5NH

RODI, W.
Imperial College,
Dept. of Mech. Eng.,
Exhibition Road, London, SW7

ROTHWELL, I. H.
Disa,
116 College Road,
Harrow, Middx, HA1 1HQ

ROYLANCE, T. F.
University of Nottingham,
Dept. of Mech. Eng.,
Nottingham, NG7 2RD

SAIY, M.
Imperial College,
Exhibition Road, London, SW7

SALAMI, Dr. L. A.
Southampton University,
Dept of Mech. Eng.,
Southampton, SO9 5NH

SCHULTZ, D. L.
Oxford University,
Engineering Department,
Parks Road, Oxford

SCHURINK, H. D.
Disa Electronik A/S,
Herlev SK 2730, Denmark

SCOTT, Dr. M. C.
Birmingham University,
Dept. of Physics,
P. O. Box 363,
Birmingham, B15 2TT

SEED, Dr. W. A.
Imperial College,
Dept of Chem. Eng.,
London, SW7

SELLBERG, L. D.
The Royal Institute of Technology,
Dept. of Aeronautics,
S-100 44 Stockholm 70, Sweden

SHAW, R.
University of Liverpool,
Dept. of Mech. Eng.,
Brownlow Hill,
Liverpool, L69 3BX

SINGH, U. K.
University of Liverpool,
Dept of Mech. Eng.,
Brownlow Hill,
Liverpool, L69 3BX

SMEDLEY, C.
University of Leeds,
Fuel Science Dept.,
Leeds, LS2 9JT

SMOLSKI, Dr. Z.
Inst. of Applied Mechanics,
Warsaw Technical University,
Warszawa, Nowoiejska 22
Poland

SOVRAN, Dr. G.
Special Programs Dept., G. M. Res.
Labs., General Motors Corp.,
G. M. Tech. Centre,
Warren, Michigan, 48090, U. S. A.

SPAVINS, G. R.
Royal Aircraft Establishment,
H. S. L. Tech. Training,
Clapham, Bedford

SPENCER, Dr. E. A.,
Head of Flow Measurement Div.,
National Engineering Laboratory,
East Kilbride, Glasgow

STASICKI, B.
ZMG, PAN, Krakow, Poland

SWITHENBANK, J.
Sheffield University,
Dept. of Chem. Eng.,
Newcastle Street,
Sheffield, S1 3JD

SYKES, D. M.
The City University,
Dept. of Aeronautics,
St. John Street,
London, EC1

SYRED, Dr. N.
Sheffield University,
Dept. of Chem. Eng. & Fuel
Technology,
Sheffield, S10 2TN

SZASZ, Dr. G.
General Electric Co.,
37 Alikanstrasse,
CH-8001 Zurich, Switzerland

TAYLOR, D. E. M.
University of Edinburgh,
Dept of Physiology,
University Medical School,
Teviot Place, Edinburgh, EH8 9AG

TERRELL, M. G.
Rolls Royce (1971)Ltd.
Sinfin A, Electronics Group,
Victory Road, Derby, DE2 8BJ

THOMAS, D. L.
C. E. G. B. Berkeley Nuclear Labs.,
Berkeley, Goucester, GL13 9PB

THOMAS, Dr. I. R.
Army School of Health, R. A. M. C.
Training Centre, Keogh Barracks,
Ash Vale, Aldershot, Hants.

THOMAS, Dr. J. R.
Gas Council,
Midlands Research Station,
Wharf Lane,
Solihull, Warwicks.

THEW, M. T.
Southampton University,
Dept of Mech. Eng.,
Southampton, SO9 5NH

TIPPING, Dr. J. C.
Gas Council,
Thermal Eng. Group,
Watson House, Peterborough Rd.,
Fulham, London, SW6

TORAZZA, A.
DTS/LAB, c/o Asgen,
Via N. Lorenzi 8,
16152 Genova-Cornigliano,
Italy

TREMEER, Dr. G. E. B.
Atomic Energy Board,
Private Bag 256,
Pretoria, South Africa

TURNER, Dr. J. T.
Manchester University,
Simon Engineering Laboratories,
Oxford Road, Manchester

TURTON, R. K.
Loughborough University,
Dept. of Mech. Eng.,
Loughborough, Leics.

TVETER, S.
University of Trondheim,
Norwegian Inst. of Technology,
N-7034 Trondheim, Nth Norway

VINCE, M.
c/o C. J. Williams,
Lysons Avenue, Ash Vale,
Aldershot, Hants.

WAD, G.
Disa Electronik A/S,
Herlev DK 2730, Denmark

WAGNER, Dr. W. J.
Max-Planck-Institut fur
Stronmungs-schung,
34 Gottingen, Bundesrepublik
Deutschland, Bottingerstr 6,
Germany

WATSON, Dr. P. J.
Cambridge University,
Engineering Department,
Trumpington Street, Cambridge

WATTS, P. R.
Fire Research Station,
of Environment,
Borehamwood, Herts.

WHAMOND, Mrs. J. S.
Edinburgh University,
University Medical School,
Dept. of Physiology,
Teviot Place, Edinburgh, EH8 9AG

WIGLEY, G.
University of Nottingham,
Dept. of Physiology &
Environmental Studies,
School of Architecture,
Sutton Bonnington,
Loughborough, Leics.

WILLIAMS, I.
University of Sheffield,
Dept of Chem. Eng. &
Fuel Technology,
Mappin Street, Sheffield, S10 2TN

WILSON, J. N.
Brighton Polytechnic,
Dept. of Eng. & Building,
Moulsecoomb, Brighton, BN2 4GJ

WIRASINGHE, N. E. A.
City University,
Dept of Mech. Eng.,
St. John St., London, EC1V 4PB

WOLFENDEN, S.
Pilkington Brothers R & D,
Applied Maths Dept.,
Lathom, Nr. Ormskirk

WOLFF, H. S.
Clinical Research Centre,
Division of Bio-Engineering,
Watford Road,
Harrow, Middx, HA1 3UJ

WRIGHT, H. E.
Air Force Institute of Technology,
Wright Patterson AFB,
Ohio, U. S. A.

YANTA, W. J.
U. S. Naval Ordnanace Laboratory,
Silver Spring,
Md. 20910, U. S. A.

AUTHORS WHO WERE NOT DELEGATES

BHINDER, F. S.
Hatfield Polytechnic,
Hatfield, Herts.

DAVIES, T. W.
University of Exeter,
Dept. of Chem. Eng.,
North Park Road, Exeter, EX4 4QF

DeSANTIS, M. J.
The Singer Company,
New Jersey, U. S. A.

DHANAK, A. M.
Michigan State University,
U. S. A.

DUBNITSHEV, Yu. N.
Institute of Automation & Electrometry,
Novosibirsk, U. S. S. R.

DUBOVIK, L. Ya.
The B. E. Vedeneev All-Union Research Institute of Hydraulic Engineering,
Leningrad, U. S. S. R.

ELROD, W. C.
Air Force Institute of Technology,
Wright-Patterson, AFB,
Ohio, U. S. A.

FALSETTI, H. L.
State University of New York at Buffalo,
U. S. A.

FOSTER, P. J.
University of Sheffield,
Dept. of Chem. Eng. & Fuel Technology,
Newcastle Street,
Sheffield, S1 3JD

HACKETT, C. E.
Massachusetts Institute of Technology,
U. S. A.

HOWARD, F. B.
Southampton University,
Dept. of Mech. Eng.,
Southampton, SO9 5NH

JANALIK, J.
University of Ostrava,
Czechoslovakia

JOHNSON, R. W.
Whirlpool Corporation,
Benton Harbor, Michigan, U. S. A.

KISER, K. M.
State University of New York at Buffalo,
U. S. A.

KORONKEVITCH, V. P.
Institute of Automation & Electrometry,
Novosibirsk, U. S. S. R.

KUDASHEV, E. B.
The B. E. Vedeneev All-Union Research Institute of Hydraulic Engineering,
Leningrad, U. S. S. R.

MORRIS, W. M.
University of Edinburgh,
Dept. of Electrical Engineering,
Kings Buildings,
West Mains Road, Edinburgh

PATRICK, M. A.
University of Exeter,
Dept. of Chem. Eng.,
North Park Road, Exeter, EX4 4QF

POLYAKOV, A. F.
Institute for High Temperatures,
Moscow, U.S.S.R.

RAKOWSKY, E. L.
The Singer Company,
New Jersey, U.S.A.

RYSZ, J.
Applied Physics Laboratory,
Polish Academy of Sciences
Poland

SAUNDERS, L. J.
The Gas Council,
Watson House, Peterborough Road,
Fulham, London, SW6

SCHMOILOV, N. F.
Institute of Automation & Electrometry,
Novosibirsk, U.S.S.R.

SKLADNEV, M. F.
The B.E. Vedeneev All-Union Research Institute of Hydraulic Engineering,
Leningrad, U.S.S.R.

SMOLARSKI, A. Z.
Applied Physics Laboratory,
Polish Academy of Sciences, Poland

SOBOLEV, V. S.
Institute of Automation & Electrometry,
Novosibirsk, U.S.S.R.

STOLPOVSKI, A. A.
Institute of Automation & Electrometry,
Novosibirsk, U.S.S.R.

TANAKA, H.
Institute of Space & Aeronatuical Science,
University of Tokyo, Japan

UTKIN, E. N.
Institute of Automation & Electrometry,
Novosibirsk, U.S.S.R.

WHITELAW, J. H.
Imperial College,
Dept. of Mech. Eng.,
Exhibition Road, London, SW7

ZAKANYCZ, S.
Air Force Institute of Technology,
Wright-Patterson AFB,
Ohio, U.S.A.

Errata for Volume I

CONTENTS. For paper I.4-2 by R.Cheesewright on p. 145 read paper II.4-2.

Paper II.2-3, p. 88.

p. 92. For equation (21) read

$$\Delta\omega_u^2 = K^2(u_o')^2 \qquad (21)$$

where $(u_o')^2$ is the mean square effective turbulent velocity.

p. 93. Delete the sentence on line 19 which reads "Clearly, because .... not so." and replace with "It is easily seen that if $\Delta\omega_G$ is significant, this procedure is invalid. In this case it may be shown that $\Delta\omega_T$ merely restores the energy from the velocity fluctuations occurring within the volume; this is not true for the instantaneous velocities which will be treated later."

Paper II.4-1, p.136.

p.136. line 2. For 'spacial' read 'spatial'.

p.141. line 16 from bottom. For 'self preserved' read 'self-preserving'

p.142. line 11. For 'flat frequency' read 'poor frequency'

Paper II.4-2, p.145.

p.145. line 18 from bottom. For 'Hassen' read 'Hassan'.

p.146. line 21. For 'Brunn' read 'Bruun'.

last line. For 'sufficiencly' read 'sufficiently'

p.149. Ref.7. For 'Brun' read 'Bruun'.

Paper II.4-3. p.152.

p.153. For equation (5) read

$$\frac{\sqrt{\overline{(u')^2}}}{u} = \frac{2\sqrt{\overline{(E')^2}}}{n(E^2-a)} - E \qquad (5)$$

p.154. For equation (6) read

$$\frac{dE}{du} = \frac{2C + Bu^{-\frac{1}{2}}}{4E} \qquad (6)$$

Paper II.4-4. p.156.

p.157. line 5 from bottom. For 'ploted' read 'plotted'

p.158. line 24. For 'standart' read 'standard'

Paper II.4-6. p.163.

p.164. line 14 from bottom. For 'spacial' read 'spatial'

p.165. Table 1. Delete "eq.(2)" from column 3.

Paper II.4-7. p.167.

p.168. Ref.1. For "Hot-Anemometers" read "Hot-film Anemometers".

Paper II.4-8. p.170.

p.173. Ref.7. For "1965" read "1968".

Paper III.1. p.183. For "T.F.Roylance, University of Nottingham and D.McWilliam, Dounreay Experimental Reactor Establishment" read "D.McWilliam, Dounreay Experimental Reactor Establishment and T.F.Roylance, University of Nottingham".

Paper III.2. p.191. Page 194 should follow page 197.

Paper IV-5. p.236.

p.239. Ref. 4. For "Journal of Applied Mathematics" read "Journal of Applied Mechanics".

Paper VII-2. p.298.

p.303. Transposition of drawings above the captions Fig. 1 and Fig. 2.

Paper VIII-3. p.305.

p.306. line 20. For "$R_o^2\omega/\nu$" read "$R_o\sqrt{\omega/\nu}$"

Index of contributors

Also published by Leicester University Press

Fluid dynamic measurements in the industrial and medical environments

Volume I Conference papers
Edited by David J. Cockrell

$9\frac{3}{4} \times 6\frac{1}{4}$ ins   352 pp   diagrams, graphs, and photographs
ISBN 0 7185 1107 7   £5·00 net

£1·00 net
ISBN 0 7185 1114 X